THE BOOKS OF
THE GRAND CANYON
THE COLORADO RIVER
THE GREEN RIVER
&
THE COLORADO PLATEAU

IN THE GRAND CANYON OF THE COLORADO
Second Powell Expedition, September 1, 1872
from Stephens & Shoemaker:
IN THE FOOTSTEPS OF JOHN WESLEY POWELL

THE BOOKS OF

THE GRAND CANYON

THE COLORADO RIVER

THE GREEN RIVER

THE COLORADO PLATEAU

1953 – 2003

A Selective Bibliography

BY MIKE S. FORD

FRETWATER PRESS • FLAGSTAFF

2003

LIBRARY OF CONGRESS
CONTROL NUMBER
2003114078

ISBN
CLOTH 1-892327-09-0
PAPER 1-892327-10-4

FIRST EDITION

PRINTED IN CANADA

To the memory of
CLEMENT DAVID HELLYER
bookman, fisherman, friend,
who first suggested this project

and for
KAY,
who didn't really mind
when I kept buying
all those books

THE GRAND CANYON, THE COLORADO RIVER, THE GREEN RIVER, AND THE COLORADO PLATEAU

When I get a little money,
I buy books;
And, if any is left,
I buy food and clothes.

– Erasmus

PREFACE

IT HAS BEEN fifty years since Francis P. Farquhar compiled the slim little volume titled *The Books Of The Colorado River & The Grand Canyon: A Selective Bibliography,* which was published by Glen Dawson as one of the volumes in the Early California Travel Series. At that time Farquhar gleaned from the published literature the writers whose work represented the most significant contributions to our knowledge of the Colorado Plateau, the Colorado River, and the Grand Canyon up to 1953. This bibliography has stood the test of time for it has been used and is still being used by countless libraries and collectors to build collections and has become a very collectible item itself. W. M. Morrison Books of Austin, Texas reprinted the Farquhar in 1991. Five Quail Books of Prescott, Arizona has a new reprint with corrections from the original and Morrison editions, to be co-released with this volume.

Since 1953 a multitude of titles has been published that contain information about the rivers, the canyon, and the plateau. Therefore, it seems time to look subjectively at this very large body of later work and select those items that contribute significantly to our knowledge and appreciation.This bibliography will include several titles that describe the upper rivers because a greater number of authors are now writing about them than before 1953. More people are running the rivers now than fifty years ago and although the largest numbers float the Grand Canyon, the upper Green, the Yampa, and the San Juan are also quite popular. Riding the rivers is not the only way to get into the country along and near the rivers and their tributaries. Hiking and backpacking have become the method of choice for many people. The following selections will include very few items that are guide books for roads or trails or for running the rivers and those few are included for different reasons.

Many of the titles that have been written since Farquhar were originally published in hardback but several were first issued in paperback. Many of the hardback items were simultaneously or have been published later in paperback. Several of the later printings were done by the original publisher but many have been reprinted by a different publisher. Only books that are part of a special printing or have special bindings or have some other feature that makes them somewhat unique have been listed or pointed out in the description of the item.

In 1974 Arizona Western College Press at Yuma published *The Lower Colorado River, A Bibliography* by Richard Yates and Mary Marshall. It lists 1425 items in eleven different categories and includes every published title or article they could locate dealing with the lower river. Useful as it is as a listing, it does not give, other than the hint the title might convey, why one might want to read it. In 1981 the Grand Canyon Natural History Association published as Monograph No. 2, *Bibliography of the Grand Canyon and the Lower Colorado River 1540-1980,* compiled by Earle E. Spamer, with William J. Breed, Robert C. Euler, and Grace Keroher. This was an all inclusive listing in ten different categories but again, for the most part, no more than a list. In 1990 as Monograph No. 8 the Grand Canyon Natural History Association published *Bibliography of the Grand Canyon and the Lower Colorado River From 1540* compiled by Earle E. Spamer with George H. Billingsley, William J. Breed, Robert C. Euler, Dorothy A. House, Grace Keroher, Valerie Meyer, Richard Quartaroli, and Lawrence E. Stevens. This was a greatly expanded second edition of the 1981 publication using 12 different categories. This item was published in an edition of 1000 loose-leaf volumes, 25 microfiche sets, and 25 sets of computer diskettes. In 1993 Supplement 1

X

was published, again compiled by Earle E. Spamer and this time with the assistance of Daniel F. Cassidy and John Irwin. The supplement was also available in loose-leaf, microfiche, and diskette. The 1990 and the 1993 publications are supplied with author indexes. At this time the total listing now is over 29,000 items in thirty-three subject areas and is available only on the World Wide Web.

Spamer has made an admirable attempt to list every title that has anything to do with the Colorado River and Grand Canyon. It is an extremely useful list for the researcher and the collector, but one lifetime would not be enough for anyone to find and read every item.

Therefore it still seems appropriate to evaluate items published since the Farquhar and try to select the best writing, the best photography, the best history, the best philosophy, the best stories, and the best combinations of these – and be informative enough about the chosen ones to encourage someone to want to read them. Nineteenth century naturalist and writer Henry David Thoreau reminded us to read the good books first because we don't have time to read them all.

Lee's Summit, Missouri   MIKE S. FORD
August, 2003

xi

ACKNOWLEDGMENTS

MANY PEOPLE have been involved in what ever this book has become. First, my parents, the late Lillian and Truman Ford, who helped me develop a love of reading. Books were a very important part of our home as they read to me and I saw them reading. Many of my early friends were books and the people in them. Soon after I started my teaching career I began traveling and on one of my earlier trips I went to Grand Canyon. After this first visit I wanted to know what I was seeing and something of the history of the area. A local used and rare book shop in Kansas City, Missouri seemed a good place to look and I became acquainted with Glenn Books and Ardis Glenn, now a valued friend. The first volume I purchased was *The Colorado* by Frank Waters and several equally good titles followed. My first edition of Farquhar also came from Glenn Books and after that I was hooked.

For the present volume I owe thanks to several people. First on that list is the late David Hellyer, founder of Five Quail Books. In Dave's book room one day we were discussing Farquhar's bibliography and I casually mentioned that someone should bring that work up to date because so many good books had been written about Grand Canyon, the Colorado Plateau, and the rivers since 1953. Dave agreed and suggested I attempt this project as I was familiar with the modern literature and with the country through which all the rivers run. After several weeks I hesitantly accepted the challenge. The first person to read the original draft was James H. Knipmeyer. He made many suggestions and most are included in the final version. Jim and I have hiked and backpacked the plateau together for many years and have visited many book stores together. Daniel F. Cassidy and Diane Cassidy, the current owners of Five Quail Books, have given invaluable help. Dan Cassidy read two draft copies and, besides offering suggestions of items that might be considered for inclusion, has made me aware of later printings and special printings and bindings of many of the titles I have listed. It is possible not all of the special printings and bindings have been included, but because of Dan this information is far more complete than it would have been. Diane Cassidy has been most generous with her time and expertise in making me aware of proper form and correcting many of my errors and omissions. She also earns a big thank you for checking and correcting the index. Brad Dimock has also been very helpful in suggesting some changes in form and wording. Last I would thank my wife Kay for reading the entire manuscript, making suggestions, and most of all giving me encouragement all these months. Any exclusions or errors of fact or form that remain are totally mine. Friends can only do so much.

When I started thinking about the books that should be included, it was my feeling that each chosen selection should add new knowledge or take a different perspective from previously published materials. It was also necessary that the work be well written and be as error free as possible historically, scientifically, and geographically. In the nearly three years I have spent putting this bibliography together I have considered or read or re-read something over twelve hundred titles. It has been a very enjoyable experience. It is amazing how much one can forget about a book that was read years before. It is also amazing what the benefit of reading other books does to one's ability to appreciate and understand better the books that are being read for the first or even second time. There are certainly other books that some readers would wish were listed in this bibliography. There were several other titles that were given strong consideration, a few early choices were dropped, and others added. Several items that are included had not yet been published when I started. If I were to start over, the selections might not be exactly the same as they are now but it would still be a very admirable list of worthy books about the areas included. This has been a most enjoyable experience and it is my hope that you will be lead to titles that give you as much information and pleasure in reading as I had in reading or re-reading them. Having said all that, I owe a very big thank you to all the authors, the editors, the photographers, the illustrators, and the people who wrote the forewords and the introductions. If they had not done the fine work they did I would have had nothing to write about and this bibliography would never have come into existence.

THE BOOKS OF
THE GRAND CANYON
THE COLORADO RIVER
THE GREEN RIVER
&
THE COLORADO PLATEAU

Part I

THE SIXTEENTH, SEVENTEENTH, AND EIGHTEENTH CENTURIES

1. BRIGGS

WITHOUT NOISE OF ARMS. THE 1776 DOMÍNGUEZ-ESCALANTE SEARCH FOR A ROUTE FROM SANTA FE TO MONTEREY. By Walter Briggs with oil paintings by Wilson Hurley and a foreword by C. Gregory Crampton. Flagstaff: Northland Press. 1976. Pp. ix, 212; ills.; maps; bibliography; index.

¶ *This is not another translation of the notes and diary from the Domínguez-Escalante expedition. Rather it is an explanation of the importance of the remarkable accomplishments of this small group of explorers who traveled over two thousand miles, for the most part into unknown territory, "without noise of arms." In the first chapter the author places the expedition in historical perspective with brief commentary on Spanish exploration in the southwest from Álvar Núñez Cabeza de Vaca to Fray Francisco Garcés. The author and artist retraced the route of 1776 and excerpts from the diary are included in the author's description and discussion of the route. Ten of Wilson Hurley's outstanding paintings are interspersed throughout the text. This narrative account is a more readable and informative version of the expedition's experiences than a direct translation of the diary. A leather-bound, numbered edition of one hundred copies in slipcase and signed by the author and artist was issued at the same time.*

3

2. DOMÍNGUEZ/ESCALANTE

a

THE DOMÍNGUEZ-ESCALANTE JOURNAL. THEIR EXPEDITION THROUGH COLORADO, UTAH, ARIZONA, AND NEW MEXICO IN 1776. Translated by Fray Angelico Chavez, and edited by Ted J. Warner with prefaces by Melvin T. Smith and Frey Angelico Chavez and an introduction by Ted J. Warner. Provo: Brigham Young University Press. 1976. Pp. xix, 203; ills.; maps; glossary; bibliography.

b

THE DOMÍNGUEZ-ESCALANTE JOURNAL. THEIR EXPEDITION THROUGH COLORADO, UTAH, ARIZONA, AND NEW MEXICO IN 1776. Translated by Fray Angelico Chavez and edited by Ted J. Warner with a foreword by Robert Himmerich y Valencia, an introduction by Ted J. Warner, and a translator's note by Fray Angelico Chavez. Salt Lake City: University of Utah Press. 1995. Pp. xxii, 153; maps; glossary; bibliography.

¶ *Without doubt this is the definitive translation of the journal made by the Franciscans Frey Francisco Atanasio Domínguez and Frey Silvestre Vélez de Escalante during their entrada of 1776. This is the first translation that uses a very early copy of the original journal—the original has not, as yet, come to light. All three items listed in Farquhar #8 used later translations and some important sentences and phrases were left out. This is also the first translation that recognizes that Frey Domínguez was actually in charge of this expedition although, much like Lewis and Clark a few years later, he and Frey Escalante worked very closely together. Warner points out the incorrect use of the name of the man we refer to as Escalante. If the expedition were named correctly it would be Domínguez-Vélez, as Vélez was really his surname and Escalante the name of his hometown in Spain. Such are the pitfalls of rendering another language into English by writers who know only English. In a the journal is also printed in Spanish. In* b *there are slight modifications in the translations and greater accuracy in several of the geographic coordinates mentioned in the notes, but it does not contain the journal in Spanish.*

3. UDALL

a

TO THE INLAND EMPIRE. CORONADO AND OUR SPANISH LEGACY. By Stewart L. Udall with photographs by Jerry Jacka. Garden City: Doubleday & Company, Inc. 1987. Pp. xviii, 222; ills.; maps; index.

b

IN CORONADO'S FOOTSTEPS. By Stewart Udall with photographs by Jerry Jacka, illustrations and paintings by Bill Ahrendt, and an essay, "The Search for Chichilticale," by Emil W. Haury. Phoenix: Arizona Department of Highways. April, 1984. Pp. 47; ills.; maps. This is *Arizona Highways* Vol. 60, No. 4, April 1984, but a limited number of this issue was also bound in hardback.

c

IN CORONADO'S FOOTSTEPS. By Stewart Udall with photographs by Jerry Jacka and an essay, "The Search For Chichilticale," by Emil W. Haury. Tucson: Southwest Parks and Monuments Association. 1991. Pp. iv, 31; ills.; map; paperback.

¶ *All three of these titles cover the Coronado Expedition route including the latest research regarding just where that route may have been. "To the Inland Empire," a, has by far the most information and photographs about Coronado's search for Cibola. It includes the probable area traversed by the Spanish in their quest from Compostela, Mexico, into what is now Kansas, which the Spanish knew as the Kingdom* of Quivira. The "Arizona Highways," b, was published before Udall had completed all of his research which is included in a. Both b and c share the same text but have different quotes from modern historians and from Spanish letters and journals. All three titles have superb photographs by Jerry Jacka but for the most part each volume has different ones. Only b has the historically accurate illustrations and paintings by Bill Ahrendt.

Part II

THE NINETEENTH AND TWENTIETH CENTURIES

4. ANDERSON

LIVING AT THE EDGE. EXPLORERS, EXPLOITERS AND SETTLERS OF THE GRAND CANYON REGION. By Michael F. Anderson. Grand Canyon: Grand Canyon Association, 1998. Pp. 184; ills.; maps; end notes, index; paperback.

¶ *Anderson begins the story with brief overviews of the long history of the first residents – the Native Americans – and the Spanish entradas. He describes contact on the fringes of the canyon country by the trappers, and the work of the government surveyors, as part of the prelude. He details the importance of the railroad across northern Arizona as the key to easier access to the South Rim, and documents early mining ventures, the first settlers, and the South Rim as a tourist destination. Many of these early miners and settlers were real characters and had their own ideas about how things were going to be. Anderson gives separate coverage to exploration and tourism on the North Rim and the Arizona Strip, pointing out the influence of the Mormon settlements in southern Utah. One of the strongest features of this work is the reproduction of myriad historical photographs, many never before published.*

5. ASHLEY

THE WEST OF WILLIAM H. ASHLEY. THE INTERNATIONAL STRUGGLE FOR THE FUR TRADE OF THE MISSOURI, THE ROCKY MOUNTAINS, AND THE COLUMBIA, WITH EXPLORATIONS BEYOND THE CONTINENTAL DIVIDE, RECORDED IN THE DIARIES AND LETTERS OF WILLIAM H. ASHLEY AND HIS CONTEMPORARIES 1822-1838. Edited by Dale L. Morgan. Denver: Fred A. Rosenstock, The Old West Publishing Company. Designed and printed by Lawton Kennedy, San Francisco. 1964. Pp. liv, 341; ills.; maps, one large folding; notes book I; notes book II; a note on the documents; index.

¶ *Only a few pages of this major work, edited by one of the West's most eminent historians, pertain to the Green River. But this book leaves little if any doubt that William Ashley led the first "float trip" on any of the waters of the Colorado River system above the Grand Wash Cliffs. From the diary it would seem that he started down the Green from a point ten to fifteen miles above the mouth of the Sandy, which would put him thirty to forty miles above the present town of Green River, Wyoming. His plan was to float the river about a hundred miles, cache goods, and mark a location for what would become the site of the first rendezvous of the fur trade. From his writings it is unclear just where Ashley's river voyage ended. He may have gone all the way through Desolation and Gray Canyons but Minnie Maud Creek, about 30 miles below the mouth of the Duchesne, is probably where he quit the river. Ashley's boat would be of interest to modern boatmen as he describes its construction. "... some of the men commenced making a frame about the size and shape of a common mackinaw boat, while others were sent to procure Buffaloe Skins for a covering." The finished "canoe" as Ashley called it, was sixteen feet by seven feet and after the first day on the river they built another one as the load was too great for just one boat. Ashley was the first to describe what Powell would later name Flaming Gorge, Horseshoe*

8

Canyon, Kingfisher Canyon, Red Canyon, and many more. Ashley's description of the rapid that Powell named Ashley Falls (because he found Ashley's inscription painted on the wall) reads much like those of later river runners. William Manly, on the way to his date with history in Death Valley, had seen the inscription twenty years before Powell, and Ellsworth Kolb said it could still be made out in 1911. This is a monumental work, both in the history it includes and in its physical size. As well as the trade edition, two hundred and fifty copies were specially bound, numbered, and signed by Morgan.

6. BERKE

MARY COLTER. ARCHITECT OF THE SOUTHWEST. By Arnold Berke with color photographs by Alexander Vertikoff. New York: Princeton Architectural Press. 2002. Pp. 320; ills.; chapter notes; bibliography; illustration credits; index; acknowledgments.

¶ *This is an in-depth biography of the complete career of Mary Colter, architect and decorator for the Fred Harvey Company and the Santa Fe Railroad. This work goes into considerable detail in regard to her research methods and her special talent for using Native American imagery to achieve her goals. She was, in a large part, responsible for the image many rail travelers had of the meaning of the "Southwest" in the early part of the twentieth century. This well researched volume explains the relationship between Fred Harvey and the Santa Fe Railroad when rail travel was the way to see the West, and highlights Miss Colter's significant part in that great era. There are many historical photographs and many of Vertikof's exceptionally nice color photographs of the remaining Colter buildings and their settings.*

7. BILLINGSLEY/SPAMER/MENKES

QUEST FOR THE PILLAR OF GOLD. THE MINES & MINERS OF THE GRAND CANYON. By George H. Billingsley, Earle E. Spamer, and Dove Menkes. Grand Canyon: Grand Canyon Association, 1997. Dedicated to John Hauert Riffey. Pp. xi, 112; ills.; maps, one folding; references; appendixes; tables; index; paperback.

¶ *From the introduction: "This is a synopsis of the prospectors, their times and mines, and the minerals they found in and around Grand Canyon. This is not an interpretive narration. It is the first consolidation of far-flung data on the subject, a starting point for more focused historical treatments."*

The authors spent many years researching various county and state records, libraries, and published sources; interviewing individuals; and spent many, many hours in the field. It is a very interesting book and the first to give broad coverage to this subject.

8. BLACKBURN/WILLIAMSON

COWBOYS & CAVE DWELLERS. BASKETMAKER ARCHAEOLOGY IN UTAH'S GRAND GULCH. By Fred M. Blackburn and Ray A. Williamson. Santa Fe: School of American Research Press. 1997. Pp. 188; ills.; map; notes; sources; picture credits; index.

¶ *In the late 1800s cowboy-archaeologists went into several of southeastern Utah's many canyons including Grand Gulch to discover and remove ancient artifacts from the cliff dwellings. Many artifacts were taken out and sold; some of the collections were deposited in various museums. Records were not always kept accurately and some items were "lost" in the museums. With cooperation from the various museums, and using reverse archaeology, a group of mostly avocational archaeologists were able to trace, correctly identify, and match many of the artifacts with the sites of origin in the canyons. This process uses field notes, names and dates the early cowboy-archaeologists had written on the ruins or the canyon walls at the sites, and early photographs of artifacts at the sites. It is a fascinating piece of detective work and includes many historical photographs. The last chapter, titled "The Future of the Past," should be required reading before anyone is allowed to go into the backcountry anywhere on the Colorado Plateau.*

9. CUSHING

THE NATION OF WILLOWS. By Frank Hamilton Cushing. With a foreword by Robert C. Euler. Flagstaff: Northland Press. 1965. Pp. 75.

¶ *This volume is the first time this material, which was first published in two 1882 issues of the "Atlantic Monthly," has been published in book form. The first article, which here is chapter one, deals with Cushing's time in Zuni. He was there five years and was so accepted by them that he was not only made a member of the tribe but was even elevated to the priesthood in one of the most secret of their religious societies. From the Zuni he learned about the Havasupai. Traveling with a guide by way of the Hopi village of Oraibi, where he spent a little time, he visited the Havasupai, who, of course, were "The Nation of Willows." He was the first ethnologist to visit the Havasupai and was early enough to find them in life patterns almost untouched by other cultures. Chapter two, which was the second article, records his accurate observations. Besides the regular edition there were two hundred slipcased copies signed by Robert Euler.*

10. GRATTAN

MARY COLTER. BUILDER UPON THE RED EARTH. By Virginia L. Grattan. Flagstaff: Northland Press. 1980. Pp. x, 131; ills.; map; notes; bibliography; Mary Colter's buildings; index.

¶ *Mary Elizabeth Jane Colter was employed by the Fred Harvey Company for more than forty years. She was a decorator and an architect for the Harvey Company and as such broke ground in two different ways. During most of her career she worked in a "man's world." At Grand Canyon she was architect and/or decorator of Hopi House, the Lookout, Hermit's Rest, the Watchtower, Phantom Ranch and Bright Angel Lodge. She was also the architect for both the Men's and Women's Dormitories. This is an excellent first biography of Mary Colter. It covers her complete career but concentrates mainly on her work at Grand Canyon. She was quite a remarkable lady. Grand Canyon Natural History Association republished this title with minor corrections in 1992.*

11. HALL

SHARLOT HALL ON THE ARIZONA STRIP. THE DIARY OF A JOURNEY THROUGH NORTHERN ARIZONA IN 1911. By Sharlot M. Hall. Edited by C. Gregory Crampton. Flagstaff: Northland Press. 1975. Pp. viii, 97. map; preface; introduction; bibliography.

¶ *Jacob Hamblin is usually given credit for being the first person of European descent to "circle" Grand Canyon. Sharlot Hall was the first woman to achieve that distinction. Her companion and guide on this over two month journey was Al Doyle, who also guided Zane Grey on some of his western trips. Doyle employed his state-of-the-art "RV"—a light-weight Studebaker wagon pulled by two Arabian ponies. At this time the Arizonans living south of Grand Canyon knew little about the Arizona Strip and as Miss Hall was Territorial Historian of Arizona, her purpose was to gather information so citizens could become better informed. The first half of her diary was published in "Arizona, the New State Magazine" from 1911 into 1913. She was the first to describe for the general public the geography, resources, and development taking place on the strip. This is a very interesting piece of Arizona history. As would be expected, Dr. Crampton has done a thorough job editing the diary and has included an interesting biographical sketch of Miss Hall in the introduction.*

12. HEGEMANN

NAVAHO TRADING DAYS. By Elizabeth Compton Hegemann with photographs by the author and an introduction by Jesse L. Nusbaum. Albuquerque: The University of New Mexico Press. 1963. Pp. xi, 388; ills.; suggested reading.

¶ *After her marriage to a Grand Canyon National Park Ranger in 1925, Hegemann lived at the Canyon for about three years. In 1929 she and her second husband purchased the trading post at Shonto, Arizona and operated it for about ten years. She was an avid and accomplished amateur photographer and the book includes 318 of her photographs. With her descriptions and stories of her experiences—at the Canyon and as an Indian trader—this book provides a valuable record of this period in the history of Northern Arizona.*

13. HUGHES

a

THE STORY OF MAN AT GRAND CANYON. By J. Donald Hughes. Grand Canyon: Grand Canyon Natural History Association. 1967. Pp. 195; ills; map; chapter references; suggested reading; index.

b

IN THE HOUSE OF STONE AND LIGHT. A HUMAN HISTORY OF THE GRAND CANYON. By J. Donald Hughes. Grand Canyon: Grand Canyon Natural History Association. 1978. Pp. 137; ills.; maps; chapter references; further reading; index; paperback.

❡ *Hughes presents a good account of man's relationship with the Canyon in these two volumes, covering early exploration, mapping, and scientific study. He tells of the mining period and the people involved, and how the development of tourist facilities made the miners richer than the mines had done. Many interesting pages tell of the establishment of the Forest Reserve, then the Game Reserve, and how that idea evolved into the National Monument and finally into the National Park. Publication* b *is an expanded and updated version of* a. *A second printing of* b *was bound in hardback.*

14. ILIFF

PEOPLE OF THE BLUE WATER. MY ADVENTURES AMONG THE WALAPAI AND HAVASUPAI INDIANS. By Flora Gregg Iliff. New York: Harper & Brothers, Publishers. 1954. Pp. xii, 271; ills.

❡ *In the year 1900 at the age of 18 Iliff left Oklahoma Territory to become a teacher to the Hualapai. A transfer to the Havasupai took place early in her career and at first she was both teacher and superintendent. These were the years when both tribes were in the early stages of transition from the old ways to many of the ways of the white man. Iliff tells of the customs, the ceremonials, everyday life and many of the things she learned while living and working with the tribes. She tells with much feeling of her friendship, her understanding, and her sincere concern for these proud people long before Supai was discovered by the tourists.*

15. KNIPMEYER

Butch Cassidy Was Here. Historic Inscriptions of the Colorado Plateau. By James H. Knipmeyer. Salt Lake City: University of Utah Press, 2002. Pp. 160; ills.; maps; afterword; notes; bibliography; index; paperback.

¶ *This is a unique look at history that is written in stone and written on stone. Carving into the rock or writing on it with axle grease, paint, charcoal, lead pencil, or lead bullet, many travelers made a record of their passing. Knipmeyer has collected photographs of over sixteen hundred inscriptions. Most of the inscriptions were made prior to 1900 by explorers, trappers, soldiers, prospectors, early settlers, cowboys, and others who were passing through the country. This book has photographs of two hundred and fifty of these inscriptions. Knipmeyer tells what these people were doing in the canyon country when they left their names and often the date. Although several of the inscriptions are by people well known in the history of the Colorado Plateau, many were made by people almost unknown to history. Most of them would have remained unknown except for the record they left on the rocks. Knipmeyer has found their stories in his very thorough research.*

16. LAVENDER

COLORADO RIVER COUNTRY. By David Lavender. New York: E. P. Dutton, Inc. 1982. Pp. xiii, 238; ills.; map; notes; bibliography; index.

¶ *Although the first three chapters trace early history, most of the information in this book falls into the nineteenth and twentieth centuries. David Lavender knows the country well and has written a concise human history of the Colorado Plateau, having pulled it from many primary and secondary sources. He includes all the major players and groups in the discovery, exploration, exploitation, and early settlement of the area drained by the Colorado River. This is an important record.*

17. LINGENFELTER

STEAMBOATS ON THE COLORADO RIVER 1852-1916. By Richard E. Lingenfelter. Tucson: The University of Arizona Press. 1978. Pp. xv, 195; ills.; maps; appendixes; notes to chapters; bibliography; index.

¶ *The boats that traveled the lower river and those that made the runs from Green River, Utah to Moab and back are here. Stanton's "Hoskaninni" in Glen Canyon, the "Charles H. Spencer," parts of which are still visible at Lee's Ferry, and Tex McClatchy's "Canyon King" are here. The dredges that dug the canals and built Laguna Dam (which put the steamboats on the lower river out of business) are all here. Illustrations and photographs almost bring the old boats to life. Lingenfelter includes the people that made it all happen, and his stories about their adventures make you aware these people were not working the usual eight to five jobs. This is a great piece of research, well written. It supplies a very important chapter in the history of the river.*

18. MANGUM

GRAND CANYON–FLAGSTAFF STAGE COACH LINE. A HISTORY AND EXPLORATION GUIDE. By Richard and Sherry Mangum. Flagstaff: Hexagon Press, Inc. 1999. Pp. 100; ills.; maps; end notes; exploration guide; bibliography; index; paperback.

❡ *This book focuses on the old stage coach route, in use from 1892 through 1900. Where did it go? Who went to the Canyon by stage? Who was operating the stage and the camps? What did things look like? How can I find these places today? All these questions and many additional ones are answered in this welcome and important addition to the history of Grand Canyon.*

19. McCROSKEY

SUMMER SOJOURN TO THE GRAND CANYON. THE 1898 DIARY OF ZELLA DYSART. Edited by Mona Lange McCroskey with a foreword by Nancy Kirkpatrick Wright. Prescott: HollyBear Press. 1996. Pp. xii, 108; ills.; maps; suggested reading; paperback.

❡ *In late July of 1898 the four Dysart siblings, ages 12 to 25, left their Phoenix home in a horse drawn wagon and seven weeks later returned. In that seven weeks they visited Wickenburg, Prescott, Grand Canyon and John Hance, the Hopi villages, Walnut Canyon, Flagstaff, Montezuma's Castle, and many other places along the way. Seventeen year old Zella kept a diary of the "sojourn" and you can re-live this delightful trip with her as our guide. She was a keen observer and a very skillful writer. This is one of the most interesting records of tourist visits to the Canyon before the railroad arrived. A fresh 2003 edition supplies a new foreword and a few different photographs and maps.*

20. POLING-KEMPES

THE HARVEY GIRLS. WOMEN WHO OPENED THE WEST. By Lesley Poling-Kempes. New York: Paragon House. 1989. Pp. xviii, 252; ills.; endpaper map; conclusion; notes; appendixes; index.

¶ *Poling-Kempes covers the entire Fred Harvey operation in this biography of the Harvey Girls and you will find of special interest those sections that deal with Grand Canyon. All the early visitors arriving by train had been served meals at various stops along the way and this continued at the Canyon. The tourist had greater enjoyment during their travels because of the many different Harvey Girls and the author tells this fascinating story. Many photographs gathered from former Harvey employees are included. The title implies the Harvey Girls opened the West. They also civilized a good part of it.*

21. RICHMOND

a

COWBOYS, MINERS, PRESIDENTS & KINGS. THE STORY OF THE GRAND CANYON RAILWAY. By Al Richmond. Flagstaff: The Grand Canyon Pioneers Society, Inc. 1985. Pp. vi, 187; ills.; maps; glossary; bibliography; paperback.

b

COWBOYS, MINERS, PRESIDENTS & KINGS. THE STORY OF THE GRAND CANYON RAILWAY. By Al Richmond. Flagstaff: Grand Canyon Railway. Sponsored by the Center for Colorado Plateau Studies of Northern Arizona University. 1989, Revised Edition. Pp. vi, 230; ills.; maps; glossary; bibliography.

¶ *The first train arrived at Grand Canyon on September 17, 1901 and trains operated on a fairly regular schedule until July 30, 1968. Both* a *and* b *are excellent records of the railroad to Grand Canyon during this time. Both have many historical photographs but* b *has some added that are not in* a. *The text in* b *is a major re-write of* a, *but the most significant difference is the reactivation of the railroad. On September 17, 1989 the "new" Grand Canyon Railway brought its first train to the Canyon and* b *contains that story.*

22. SMITH

THE SOUTHWEST EXPEDITION OF JEDEDIAH S. SMITH. HIS PERSONAL ACCOUNT OF THE JOURNEY TO CALIFORNIA 1826–1827. By Jedediah S. Smith, edited with an introduction by George R. Brooks. Glendale: The Arthur H. Clark Company. 1977. Pp. 259; ills.; maps; bibliography; index.

¶ *Not many pages of this volume are concerned with the Colorado River but it seems worthy of a place in this list. This is the journal, discovered in 1967 in an old St. Louis home, of the leader of the first American party to cross the Lower Colorado River and enter California by land. Smith and his party started their trek from the 1826 rendezvous at Soda Springs just north of Bear Lake in what is now Idaho. They journeyed generally a little west of south and eventually struck what he called the Adams River, which we now know as the Virgin. Following this stream to the large river his native guides had told him about, Smith wrote "… it could be no other but the Colorado of the west which in the Mountains we call seetes-ker-der." [The Mountain Men knew this early!] Crossing the Colorado near the mouth of the Virgin he went inland a little to the east to avoid the canyon his guides told him about. Eventually working back to the river, he went south to the Mojave Villages and crossed into California. Smith was an excellent observer and a good writer and must have had a built-in Global Positioning System. Editor Brooks has done an outstanding job and the footnotes keep you well aware of where Smith and his party were located. The journal brings Smith, but not all of his party, back to the 1827 rendezvous. For almost the entire route, Smith was breaking new trail.*

23. SWEENY

JOURNAL OF LT. THOMAS W. SWEENY 1849-1853. Edited and with an introduction by Arthur Woodward. Los Angeles: Westernlore Press. 1956. Pp. ix, 278; ills.; bibliographical notes; index.

¶ *Late in 1850 Lt. Sweeny accompanied Major Heintzelman with three companies from San Diego to the mouth of the Gila—their goal to establish a post for the protection of immigrants on their way to California. Sweeny's relationship with Heintzelman reminds one of Lt. Cave Couts relationship with Lt. Whipple [Farquhar #14 and #18]. Sweeny was in the Mexican War and was wounded twice; the second time was serious and he lost his right arm. In the Civil War he was in the battle at Shiloh where he again received a serious wound but recovered in time to be wounded one additional time before the war ended. His experiences at Camp Yuma are written in a bold and lively manner and his criticism of his senior officer is not disguised. Sweeny established a good relationship with the local Indians and the immigrants spoke very well of him. A large portion of this journal was published in "The New York Atlas" from December 1856 through March 1857 under the title "Life on the American Desert" and an edited version was printed in "The Journal of American Military Institution" in 1909. The present editor has used these two sources to put together as near an original version of Sweeny's Journal as is possible.*

24. WAY

a

A Summary of Travel to Grand Canyon. By Thomas E. Way. Prescott: Prescott Graphics. 1980. Pp. 14; ills.; maps; footnotes; paperback.

b

Destination: Grand Canyon. The Story of Travel to the Grand Canyon. By Thomas E. Way. Phoenix: Golden West Publishers. 1990. Pp. 112; ills.; maps; notes; bibliography/reading list; index; paperback.

¶ *Both of these titles cover the early days of travel and visits to Grand Canyon. Way describes tourist accommodations, those who operated them, and the merchants who provided services. The text in* b *is expanded both in scope and depth over* a. *Both have several interesting historic photographs; several are common to both.*

25. WEBER

The Taos Trappers. The Fur Trade in the Far Southwest, 1540-1846. By David J. Weber. Norman: University of Oklahoma Press. 1971. [1968 on microfilm]. Pp. xiii, 263; ills.; maps; bibliography; index.

¶ *There is information here not found in Robert Glass Cleland's "This Reckless Breed of Men," Farquhar #12. With the use of Mexican archives the author clarified some previous information and corrected some errors. The trapper could get supplies in Taos and sometimes sell their furs there, depending on the mood of the Mexican authorities. The trails from Taos took the trappers to the Gila, the San Juan, the Colorado, the Grand, the Gunnison, the Green, and many of the other tributaries of the Colorado. Many of the trappers' names are familiar. Ashley is here as are Ewing Young,*

James Ohio Pattie, Jedediah Smith, "Peg-leg" Smith, the Robidoux brothers, George Yount, and Kit Carson. You will want both this book and the Cleland to get the full picture of the importance of the trappers—the real "pathfinders" of the West.

PART III

THE MORMONS ON THE COLORADO PLATEAU

26. BROOKS

a

JOHN DOYLE LEE. ZEALOT – PIONEER BUILDER – SCAPEGOAT. By Juanita Brooks. Glendale: The Arthur H. Clark Company. 1961. Pp. 404; ills.; maps; appendix; bibliographical note; index.

b

JOHN DOYLE LEE. ZEALOT – PIONEER BUILDER – SCAPEGOAT. By Juanita Brooks. Glendale: The Arthur H. Clark Company. 1972, second edition with corrections. Pp. 404; maps; appendix; bibliographical note; index.

¶ *Just two chapters in this book relate directly to the Colorado River but those are of prime importance. First, Lee built and operated the first ferry at the river crossing place we now call Lee's Ferry. Second, the Mormon Church and its members played a significant role in exploration and settlement of much of the Colorado Plateau. Therefore it seems appropriate to follow that history through one of the better known members of the church. Brooks picks up this history with Lee joining the church near the settlement of Far West in western Missouri. She then follows him to Utah, to the ferry, and finally to his trial and execution for his part in the Mountain Meadows Massacre. Many of his activities did not differ greatly from those of the people he associated with, but in many instances Lee's role had greater visibility. This was a hard country to live in and Brooks brings the story of these determined people to life. Based on new information, there are a few additions and corrections in* b.

27. COX/RUSSELL

FOOTPRINTS ON THE ARIZONA STRIP. By Nellie Iverson Cox and assisted by Helen Bundy Russell. Bountiful: Horizon Publishers. 1973. Pp. 256; ills.; map; index.

¶ *There is much here about settlement and life on the Arizona Strip that is difficult or impossible to find in any other source. The history here is centered around Mt. Trumbull or, as it was referred by the people who lived there, Bundyville. The authors are descendants of early settlers of the Bundyville area and much of the information in the book has come down as family history. A second edition was published in 1982 and a third edition in 1998. Each is by a different publisher but all three editions share the same text. Mrs. Cox has written another book titled "The Arizona Strip – A Harsh Land and Proud." Published in 1982, it continues the history of the Strip.*

28. CREER

MORMON TOWNS IN THE REGION OF THE COLORADO, AND THE ACTIVITIES OF JACOB HAMBLIN IN THE REGION OF THE COLORADO. By Leland Hargrave Creer, edited by Robert Anderson. Salt Lake City: University of Utah Press. May, 1958. University of Utah Department of Anthropology, Anthropological Papers Numbers 32 and 33, Glen Canyon Series Numbers 3 and 4. Pp. iv, 35; map; bibliographies; paperback.

¶ *The first part of this book is a history of the settlement and the settlers of the small towns and villages along or near the Colorado River and the tributary streams. The settlements in south-eastern Utah and north-eastern Arizona included are Boulder, Cannonville, Escalante, Hanksville, Hite and Lee's Ferry. These were the last of the settlements and were in the most rugged and remote areas. They all have very interesting stories. The second paper gives information about Jacob Hamblin's work in settlement, peacemaking, and assisting Major Powell.*

29. GEARY

THE PROPER EDGE OF SKY. THE HIGH PLATEAU COUNTRY OF UTAH. By Edward A. Geary. Salt Lake City: University of Utah Press. 1992. Pp. 282; ills.; map; notes; index; paperback.

¶ *This is as near a sequel to Dutton's "Tertiary History" [Farquhar #73] as we will probably ever have. It is not a book of geology but rather a book about what the people who have settled in the high plateau country have been able to do – or not do – because of the geology. There is a lot of history here and all of it tied to the landforms and their effect on the activities of the people. Wallace Stegner's "Mormon Country" should be on a nearby shelf.*

30. MILLER

a

HOLE IN THE ROCK. AN EPIC IN THE COLONIZATION OF THE GREAT AMERICAN WEST. By David E. Miller. Salt Lake City: University of Utah Press. 1959. Pp. xi, 229; preface; ills.; end-paper maps; maps; appendixes; bibliography; index.

b

HOLE IN THE ROCK. AN EPIC IN THE COLONIZATION OF
THE GREAT AMERICAN WEST. By David E. Miller. Salt
Lake City: University of Utah Press. 1966, second edition.
Pp. xiv, 229; preface to the first edition; preface to the
second edition; ills.; end-paper maps; maps; appendixes;
bibliography; index.

¶ *These people didn't know it was impossible to build a wagon
road from the little town of Escalante, Utah to Montezuma
Creek on the San Juan River—so they went ahead and did it.
They traveled as far as what is now the town of Bluff, Utah,
less than twenty miles short of their destination. Don't call
them failures until you have covered all of the ground. It
doesn't matter what kind of vehicle you have, you won't make
it! This may be the only book you need to read about the San
Juan Mission and the Hole-in-the-Rock route of 1879-80.
David Miller covered all the ground, most of it more than
once. He has used all the journals and letters that were
available and these can be found in the appendixes along
with the list of the personnel of the expedition. These two
publications are nearly the same but* b *does contain a small
amount of material discovered since the publication of* a, *the
most significant being the brief journal of Silas S. Smith,
President of the San Juan Mission.*

31. PETERSON

TAKE UP YOUR MISSION. MORMON COLONIZING ALONG
THE LITTLE COLORADO RIVER 1870-1900. By Charles S.
Peterson. Tucson: The University of Arizona Press. 1973.
Pp. xii, 309; ills.; endpaper maps; maps; bibliography;
index.

¶ *There were three trails from Utah to the Little Colorado
country and all were difficult. Good water was scarce for*

26

much of the way. The northern route crossed the Colorado at Moab, and the San Juan at Montezuma Creek. The southern route crossed the Colorado at either Stone's Ferry or Pearce Ferry, below Grand Canyon. The most direct and popular route crossed the Colorado at Lee's Ferry. Peterson is a former director of the Utah State Historical Society and a descendant of colonists who participated in this colonization. He has used unpublished diaries and personal records, and all the published materials available regarding what eventually became a successful colonization.

32. REILLY

LEE'S FERRY. FROM MORMON CROSSING TO NATIONAL PARK. By P. T. Reilly. Edited by Robert H. Webb with contributions and epilogue by Richard D. Quartaroli. Logan: Utah State University Press. 1999. Pp. xvi, 542; ills.; maps; epilogue; chapter notes; appendixes; index.

¶ *This may not be but probably should be the last book anyone does on Lee's Ferry. Reilly didn't leave much of anything new for the next writer. He spent the better part of fifty years gathering and sorting information. He includes all the facets and phases of the people, and the history they were part of, at or near this isolated crossing place on the Colorado River. The first visit by Domínguez and Escalante is included. The very complex life of John D. Lee, the Mormon settlers' crossings to go southeast to the Little Colorado, and Jacob Hamblin's many crossings are covered. John Wesley Powell and the later river explorers, the prospectors, and the modern river runners are all here. Everything of any importance that ever happened at the place we call Lee's Ferry is in this long awaited treatise. It is unfortunate Reilly did not live to see the completed book.*

33. RUSHO/CRAMPTON

a

DESERT RIVER CROSSING. HISTORIC LEE'S FERRY ON THE COLORADO RIVER. By W. L. Rusho and C. Gregory Crampton, with a foreword by Senator Barry M. Goldwater. Salt Lake City and Santa Barbara: Peregrine Smith, Inc. 1975. Pp. 126; ills.; maps; tour section; reading list; footnotes [endnotes]; index; paperback.

b

LEE'S FERRY. DESERT RIVER CROSSING. By W. L. Rusho and C. Gregory Crampton with the foreword to the 1975 edition by Senator Barry M. Goldwater. Salt Lake City: Cricket Productions. 1992. Pp. 168; ills.; maps; tour section; reading list; endnotes; index; paperback.

c

LEE'S FERRY. DESERT RIVER CROSSING. By W. L. Rusho and with contributions by C. Gregory Crampton and the foreword to the 1975 edition by Senator Barry M. Goldwater. Salt Lake City: Tower Productions. 2003. Pp. 244; ills.; maps; endnotes; tour section; reading list; index; afterword; paperback.

¶ *These titles, by two writers very familiar with the history and geography of the Lee's Ferry area, will satisfy the needs of most people in their quest for knowledge about this historic desert river crossing. The tour sections are very helpful for those that wish to spend a few days exploring.*

There was a revised second printing of a *in 1981. Although* b *has a different title it is basically the same as* a *but in a slightly larger format. In 1998* b *was reprinted by Tower Productions of Salt Lake City and St. George with more pages due to larger type. Again as in* b, c *has many more pages but mainly because of larger type. Each succeeding printing has some additions and revisions and the most noticeable difference is in the quality of the photographs which are much better in* c. *The afterword in* c *is a four page tribute to Dr. C. Gregory Crampton.*

Part IV

DOWN THE RIVERS AND
THROUGH THE CANYONS

34. BAKER

TRAIL ON THE WATER. By Pearl Baker with an introduction by O. Dock Marston. Boulder: Pruett Publishing Company, 1969. Pp. 134.; ills.; front end sheet map and five maps on two sheets in rear pocket.

¶ *Albert [Bert] Loper was involved in many activities at various times and places on the Colorado River. In the fall of 1907 Charles Russell, Edwin Monett, and Loper left Hite with the intention of going all the way through Grand Canyon on a prospecting trip. Because he had to wait to get a camera shutter repaired, Loper was left behind. After that he worked some gold claims in Glen Canyon for several years, settling at the mouth of Red Canyon. He visited with the Kolb Brothers as they went through Glen Canyon in 1911. He helped bury Cass Hite at Ticaboo. He was head boatman on the 1921 survey of the San Juan River and the 1922 survey of the Green River. In 1939, with Don Harris, he made his long awaited trip through Grand Canyon. In the 1940s he guided many Boy Scout trips through Glen Canyon. In 1949, to celebrate his eightieth birthday, again with Don Harris and several others, he started a trip down Grand Canyon. At 24½ Mile Rapid Loper's boat capsized and he disappeared. The party found the boat but did not recover Loper's body. What were probably his bones were found in 1975 and buried in Salt Lake City. [See Ronald Ives article in "The Journal of*

Arizona History," Spring 1976.] Most people think he would
have rather stayed in the Canyon. Using Loper's diary and
personal interviews the author, who knew him all her life, has
given us a good – if sometimes fanciful – picture of the "grand
old man" of the Colorado.

35. BEER

WE SWAM THE GRAND CANYON. By Bill Beer. Seattle: The
Mountaineers. 1988. Pp. 171; maps; ills.

¶ *The sub-title on the dust jacket of this book is "The True*
Story of a Cheap Vacation That Got a Little Out of Hand."
That may be one of the great understatements of all time.
In early April 1955, Beer and good friend John Daggett
waded into the red/brown fifty-one degree water at Lee's
Ferry wearing swim fins, wool long johns and thin rubber
shirts. Their food, cooking gear, warm clothes, and any other
supplies were stored in war surplus rubber boxes which also,
along with life vests, provided flotation. Twenty-six days later
they came out of the Canyon at Pearce Ferry, much thinner
and much wiser. It is a fascinating story of adventure that no
one will ever do again. "Collier's Magazine," August 5, 1955
had a very short version of this story.

36. BLAUSTEIN/ABBEY

THE HIDDEN CANYON. A RIVER JOURNEY. Photographs and
preface by John Blaustein, A Journal by Edward Abbey,
and an introduction by Martin Litton. New York: The
Viking Press, 1977. 135 Pp.; ills.; map; notes to the plates.

¶ *Boatman/photographer Blaustein rowed over thirty eighteen-day dory trips through the Canyon with passengers on board to make the photographic record in this book. A great variety of photographs are included and several are taken from the dory as he takes it through a rapid. He used a special waterproof rig to hold the camera so he could row with both hands, thank goodness! Abbey's journal describes, as only Abbey could, the trip down the river. He also includes some appropriate quotes from John Wesley Powell. Chronicle Books of San Francisco reprinted this title in both hardback and paperback in 1999. Both publications are essentially the same. In the later printing there are some additions to the Litton introduction mainly because of the death of Edward Abbey between the two dates of publication. There are also a few changes in the order, selection and size of the images. No doubt due to improvements in the technology for printing color photographs some of the images in the 1999 printing may have been reproduced with a slightly truer color but in both books they are excellent. If you have been down the river through Grand Canyon this book will bring back many vivid memories—mostly pleasant ones but an occasional "What am I doing here?" If you haven't been down the river it will give you a hint of what you have missed.*

37. CALVIN

THE RIVER THAT FLOWS UPHILL. A JOURNEY FROM THE BIG BANG TO THE BIG BRAIN. By William H. Calvin. New York: Macmillan Publishing Company. 1986. Pp. xii, 528; maps; ills; end notes and bibliography.

¶ *To tell his story Calvin, a neurobiologist, combined four Grand Canyon river trips into one. [One early river runner has been criticized for doing something like that!] Evolution, especially of the "big brain," man's brain, is discussed in relation to the evolution of the Canyon. Many appropriate and very interesting quotations from various authors are placed in the text margins. Calvin's expertise, along with that of his traveling companions, makes this both an entertaining and a very stimulating book. Another interesting title by the same author is "How the Shaman Stole the Moon. In Search of Ancient Prophet-Scientists from Stonehenge to the Grand Canyon."*

38. COOK

THE WEN, THE BOTANY, AND THE MEXICAN HAT. THE ADVENTURES OF THE FIRST WOMEN THROUGH GRAND CANYON, ON THE NEVILLS EXPEDITION. By William Cook. Orangevale: Callisto Books. 1987. Pp. vi, 151; ills.; map; paperback.

¶ *Relying heavily on river diaries and interviews of participants plus material provided by both of Nevills's daughters, Cook has put together a complete picture of this historic river trip through Grand Canyon. This was Nevills's first voyage through Grand Canyon. Dr. Elzada Clover and Lois Jotter, the first women to go down the Colorado through Grand Canyon, were among his passengers. This is an exciting telling of an early commercial river running experience when the people in charge were just learning about the things that had to be done for a successful trip. Commercial river running has changed a great deal since that time.*

39. CRUMBO

A RIVER RUNNER'S GUIDE TO THE HISTORY OF THE GRAND CANYON. By Kim Crumbo with a foreword by Edward Abbey. Boulder: Johnson Books. 1981. Pp. 55; ills.; maps; bibliography; paperback.

¶ *Crumbo has worked at Grand Canyon as a river guide and as a river ranger. In this book, designed to fit nicely in an "ammo can," he has discussed significant historical river-related information and given the river-mile for the event. There are later printings with a revision in Abbey's foreword. It seems the park service did not approve of his suggestion for some of the equipment river runners might carry to help solve the over flight situation.*

40. DAWDY

GEORGE MONTAGUE WHEELER. THE MAN AND THE MYTH. By Doris Ostrander Dawdy. Athens: Swallow Press/Ohio University Press. 1993. Pp. viii, 122; maps; appendixes; notes; selected bibliography; index.

¶ *All students of western history know George M. Wheeler as the leader of one of the four "Great Surveys," but until this biography was published most did not realize what Wheeler was really looking for in the West. It seems his real goal was to find out where the best areas were for locating rich mineral claims and then forming mining companies to eventually mine those claims. Dawdy brings out Wheeler's unethical treatment of his men in this long overdue addition to the history of the Colorado River and the Colorado Plateau.*

41. DIMOCK

SUNK WITHOUT A SOUND. THE TRAGIC COLORADO RIVER HONEYMOON OF GLEN AND BESSIE HYDE. By Brad Dimock. Flagstaff: Fretwater Press, 2001. Pp. xii, 288; end paper maps; glossary; photo credits; index.

¶ *On October 20, 1928, Glen and Bessie Hyde launched their scow at Green River, Utah. This was a delayed* honeymoon trip and the goal was to float the Green River to the Confluence and follow the Colorado River all the way through Grand Canyon. Author Dimock and his wife, Jeri Ledbetter, are both professional river guides with many Grand Canyon trips behind them. However, one of those trips will probably stand out in their memories above all others. As part of the research for this book they built a scow, as much like the Hydes' as possible, and floated from Lee's Ferry through the Canyon. After several harrowing experiences in the rapids they developed a great deal of respect for Glen Hyde's ability to handle his scow. Dimock has conducted a thorough research, consulting all the sources available including family records, and has written a story that dispels all the myths. It will probably never be known for sure just what happened to the Hydes but this book comes as close as anyone can unless new material is discovered.

42. DOLNICK

DOWN THE GREAT UNKNOWN. JOHN WESLEY POWELL'S 1869 JOURNEY OF DISCOVERY AND TRAGEDY THROUGH THE GRAND CANYON. By Edward Dolnick. New York: HarperCollinsPublishers. 2001. Pp. 367; maps; end paper maps; ills; notes; bibliography; index.

¶ A 'New York Tribune' reporter came to Powell when the Brown-Stanton expedition was aborted after losing three men and asked him how he had made it down the river. Powell replied, "I was lucky." Dolnick has used all the known journals, diaries, letters, newspaper articles, interviews, and any other information available from all the members of Powell's 1869 journey. He has written a new and more complete "journal" of the expedition than any one member could have done. He brings into the story other historical events that had important influences on the men such as the Civil War, and for Powell, Shiloh and the Hornet's Nest. He points out how some of those events may have caused them to think and do some of the things they did. He puts all the events of the 1869 trip into historical perspective with the complete story and when you finish you almost have a feeling you've never read this story before. Yes, Powell was lucky but he made some decisions that put the odds in his favor. There is a London edition published in 2002. There are some minor changes in the text in the London edition and it has 452 pages due mainly to format size. The New York edition is bound in pictorial boards with a half-jacket; the London edition is in cloth with a standard dust jacket.

43. FLAVELL

THE LOG OF THE PANTHON. AN ACCOUNT OF AN 1896 VOYAGE FROM GREEN RIVER, WYOMING TO YUMA, ARIZONA THROUGH THE GRAND CANYON. By George F. Flavell, edited by Neil B. Carmony and David E. Brown with a foreword by Barry Goldwater. Boulder: Pruett Publishing. 1987. Pp. ix, 109; ills.; maps; selected bibliography; index; paperback.

¶ *From the introduction: "…the 'Log of the Panthon' is not a collection of cursory notes and remarks – it is a literate narrative written in a lively style with considerable humor." Of interest also is the fact that it was written while on the river, not edited later as is the case of most river diaries and journals. Flavell, with his companion, Ramon Montez, started from Green River, Wyoming on August 27, 1896, arriving in Yuma January 8, 1897, "a poorer but wiser man." Also it seems Flavell and Montez were very good at running rapids as the only rapid they portaged or lined between Lee's Ferry and Yuma was Soap Creek. They were the first to run Lava Falls which was not known to be run again until 1928. This little book is a great late addition to the river running literature of Grand Canyon.*

44. FLETCHER

RIVER. ONE MAN'S JOURNEY DOWN THE COLORADO, SOURCE TO SEA. By Colin Fletcher. New York: Alfred A. Knopf. 1997. Pp. 400; ills.; maps; garlands and accolades including sources quoted or leaned on.

¶ *In his book "The Thousand Mile Summer" Fletcher wrote about walking the length of California. In "The Man Who Walked Through Time" he wrote about walking the length of Grand Canyon National Park, as it was then. After those two trips Colin Fletcher couldn't be expected to do anything less than completely. So, if he said he was going to follow the Green River to the Colorado River and then to the Gulf, you would expect him to start as high in the mountains as he could. That is exactly what he did. He backpacked to Green River Pass and found the highest water source he could reach in mid July. He followed that growing mountain stream down to the Green River Lakes, got into his raft, and went all the way to salt water. It was one of the few years since the dams that the river ran that far. Fletcher's writing style lets him share his many experiences with you in a very personal way. It is an experience few in their late sixties would care to duplicate.*

45. FOWLER

THE WESTERN PHOTOGRAPHS OF JOHN K. HILLERS. MYSELF IN THE WATER. By Don D. Fowler. Washington: Smithsonian Institution Press. 1989. Pp. 166; ills.; maps; chapter notes; appendix; bibliography.

¶ *John K. Hillers became a photographer by chance. Hillers joined Powell on the second river trip in 1871 and stayed with him until 1900. He started out as a boatman but showed an interest in photography early in the trip and worked with E. O. Beaman as much as possible. Early in 1872 Beaman was discharged. James Fennemore was hired as photographer and Hillers continued as an assistant. Fennemore became ill at Lee's Ferry and left the party before the expedition started down Marble Canyon. Hillers became the photographer. Every scene he exposed from Lee's Ferry to and up Kanab Canyon was a "first." Although many of Hillers' photographs in this book are not of Grand Canyon or even on the Colorado Plateau, the collection printed here shows how skilled Hillers was as a landscape and portrait photographer. His nearly thirty years with Powell made their association the longest of any of Powell's men.*

46. GHIGLIERI

CANYON. By Michael P. Ghiglieri. Tucson: The University of Arizona Press. 1992. Pp. xvi, 311; maps; epilogue; selected references; index.

¶ *Wow! What a ride! Ghiglieri has a Ph.D. in biological ecology and has been a professional river guide for many years. His descriptions of running the river do about everything being there would do except get you wet. He also brings into his writing the effect Glen Canyon Dam and the Bureau of Reclamation policies have had on Grand Canyon. A great amount of river history is brought into the story as the river is run.*

47. GHIGLIERI

FIRST THROUGH GRAND CANYON. THE SECRET JOURNALS AND LETTERS OF THE 1869 CREW WHO EXPLORED THE GREEN AND COLORADO RIVERS. By Michael P. Ghiglieri with writings by George Young Bradley, Andrew Hall, William Robert Wesley Hawkins, Oramel G. Howland, John Wesley Powell, Walter H. Powell, and John Colton

38

Sumner, and with a foreword by Richard D. Quartaroli. Flagstaff: Puma Press. 2003. Pp. xvii, 342; ills.; maps; epilogue; appendix; sources; index.

¶ *Ghiglieri has gone back to the original writings of the members of the Powell expedition of 1869 and transcribed the faded letters and water-stained journals word for word. This transcription includes a few pages, paragraphs, and sentences that had been transcribed incorrectly or omitted in previously published materials. He has also included excellent biographical material about each member of the expedition. One of the several strengths of the book is the treatment of the possible fates of William Dunn and the brothers Seneca and Oramel Howland, the three who left Powell at what was later named Separation Rapid. There is much food for thought as you read this one.*

48. GOLDWATER

a

A JOURNEY DOWN THE GREEN AND COLORADO RIVERS 1940. By Barry M. Goldwater. Phoenix: H. Walker Publishing Company. 1940. Pp. 106 mimeographed on one side; map; stapled gray paperback.

b

AN ODYSSEY OF THE GREEN AND COLORADO RIVERS. THE INTIMATE JOURNAL OF THREE BOATS AND NINE PEOPLE ON A TRIP DOWN TWO RIVERS. By Barry Goldwater. Privately printed, March 3, 1941. Pp. np. [32]; map; ills.; paperback.

c

DELIGHTFUL JOURNEY. DOWN THE GREEN & COLORADO RIVERS. By Barry M. Goldwater. Photographs by the author. With supplemental essays: "Prehistoric Man in the Grand Canyon" by Robert C. Euler and "Geological Review of the Colorado Canyons" by Carlton B. Moore. Special Consultant, O. Dock Marston. Tempe: Arizona Historical Foundation. 1970. Pp. 209; maps; ills.; suggestions for further reading.

¶ *Barry Goldwater made his first trip down the Green and Colorado Rivers in 1940 with Norman Nevills. He was in "The First Hundred" to go through Grand Canyon and the complete listing is given near the back of* c. *In all three versions the author brings in some river history and gives a good account of Norm Nevills's operation of river trips. It is the best journal of commercial river running prior to the construction of Glen Canyon Dam. The text in* c *is a slightly expanded and modified version of the diary the author kept on the river. Both* b *and* c *have many photographs most of which were taken by the author. In* c *the photographs are placed as near as possible to the appropriate text. Several pages of* c *include quotes from John Wesley Powell to whom the book is dedicated. A "greatly condensed version" of the diary,* a, *was printed in* b *and this small edition was in print only a short time. In* b *the photographs are not in the order of the trip and many but not all are included in* c. *A small portion of the text was also adapted for an article that appeared in the January 1941 "Arizona Highways" magazine. The river diary,* a, *was printed in an edition of three hundred copies and was distributed privately. All of* a *was hand numbered and signed by the author. "Delightful Journey," was also issued in a specially bound, slipcased, and signed edition of one hundred copies with an original photograph laid in.*

49. HAMILTON

WHITE WATER. THE COLORADO JET BOAT EXPEDITION 1960. By Joyce Hamilton. Christchurch: The Caxton Press. 1963. Pp. 259; ills.; endpaper maps; appendix.

¶ *Most river runners feel enough challenge just getting a boat safely down the river. This is a story about a group of New Zealanders and Americans who attempted an up-river run and finally conquered the pre-Glen Canyon Dam Colorado with jet boats. The author is the wife of Jon Hamilton, son of the jet-boat inventor. Jon Hamilton was the main pilot on the successful up-river run. He piloted each of the four boats through Vulcan Rapid [Lava Falls]. "Dock" Marston and Bill Belknap were in the crew. It is an exciting story of a trip that will never be done again. Also of interest by the same author is a little thirty-page booklet titled "Diary Kept During the Upriver Conquest of the Colorado River, June-July, 1960." The author was on the downstream river trip which cached fuel and during the upriver run she and others ran the rims to keep visual contact, when possible, with the people on the river.*

50. HENRY

ROW AWAY FROM THE ROCKS. By George T. Henry with photographs by the author. Cedar Rapids: George T. Henry. 1992. Pp. vii, 90. ills.; foreword; epilogue; paperback.

¶ *The title gives good advice no matter what river you are on. This volume is about river trips, mostly on the Yampa and the Upper Green. Included are trips through the Canyon of Lodore, Whirlpool Canyon, Island Park, and Split Mountain Canyon in Dinosaur National Monument. Henry is a professional photographer turned professional boatman. He started his boating career with Bus Hatch and has lots of stories about river running beginning with the late '50s and early '60s when commercial raft trips were just becoming popular. Henry mentions some Salmon River trips as well as a trip through Glen Canyon as the reservoir was filling. There is also a limited hard cover printing.*

51. HILLERS

"Photographed All the Best Scenery." Jack Hillers's Diary of the Powell Expeditions, 1871-1875. Edited by Don D. Fowler. Salt Lake City: University of Utah Press. 1972. Pp. 225; ills.; epilogue; Jack Hillers's Photographs of the Powell Expeditions, 1871-1875, index of personal names; index of place names.

¶ *When the Utah Historical Society printed the journals and diaries related to the Powell expeditions [Farquhar #47, 48], the Hillers diary was still in the possession of the family. Since that time it has become the property of the Smithsonian National Anthropological Archives. The diary begins May 16, 1871 and ends June 10, 1875. Some of the entries are just one line but many go into detail about that day's activities. There are no entries for several days, up to about three weeks at one point, but altogether the diary gives a very good idea of what Hillers and the rest of the party were doing. He was an excellent observer of what was going on and that trait carried over into his photographs. Forty-four photographs are reproduced, most of which are by Hillers. Although some historians and researchers believe several photographs credited to Hillers are really the work of Beaman, that controversy has little or no bearing on the value of this book. Fowler's work in the introduction and notes leaves very few unanswered questions for the reader.*

52. HOLMSTROM

Every Rapid Speaks Plainly. The Salmon, Green & Colorado River Journals of Buzz Holmstrom Including the 1938 Accounts of Amos Burg, Philip Lundstrom, and Willis Johnson. Edited by Brad Dimock, with preface and introduction by the editor. Flagstaff: Fretwater Press. 2003. Pp. xxviii, 252; ills.; maps; notes; index; paperback.

¶ *The editor of this volume was one of three co-authors of "The Doing of the Thing. The Brief Brilliant White Water Career of Buzz Holmstrom," #71 in this list, the complete biography of Buzz Holmstrom. In researching the material for the biography, Dimock found Holmstrom's river journals to be essential sources. Now we have complete and accurate transcriptions of all the Holmstrom journals for three rivers, along with the journals of the three people who joined him on the 1938 trip. The Salmon River trip in 1936 was one of the first times anyone had used a row boat on the Salmon or Snake Rivers. Holmstrom learned a lot on that trip. The 1938 trip with Amos Burg, who took the first rubber raft down the Green and Colorado Rivers, is a very interesting trip. Holmstrom was alone in 1937 and the journal of his trip from Wyoming until he bumped the bow of his boat up against Boulder Dam is a true joy to read. He was one of the best—maybe the very best—boatman to ever run the Green and Colorado. He was also the most humble. Two hundred copies were also printed in hardback, numbered, signed by the editor, and in slipcase with "The Brave Ones," #54 in this list. These two books are the first volumes of "The Colorado River Chronicles" series.*

53. KNOWLTON

RIVER LOVE. THE COLORADO RIVER FROM THE ROCKIES 1,450 MILES TO MEXICO. By Smokey Knowlton. La Habra: Robert Valentine Knowlton. 1985. Pp. 112; maps; ills.; acknowledgments; paperback.

¶ *This is a unique book and a unique story about a father and three teenagers—a daughter, a son, and the son's friend—who floated the Colorado from high in the state of Colorado to the Confluence, then all the way to where the river sank in the sand near the Gulf. Based on their previous lack of experience and the boat used, they probably shouldn't have made it. They took many photographs on the journey. Although many of them are not of the best quality they show what an exciting and fun trip this was. The text is from diaries and tapes. With notes on the sketch maps the reader is able to keep up with the trip. Knowlton had a dream of doing something with his family and they all worked and played together to make that dream come true. You will share their pride after finishing the book.*

54. KOLB

THE BRAVE ONES. THE JOURNALS AND LETTERS OF THE 1911-1912 EXPEDITION DOWN THE GREEN AND COLORADO RIVERS. BY ELLSWORTH L. KOLB AND EMERY C. KOLB, INCLUDING THE JOURNAL OF HUBERT R. LAUZON. Transcribed and edited by William C. Suran and with a foreword by Brad Dimock. Flagstaff: Fretwater Press. 2003. Pp. xix, 180; ills.; maps; endnotes; photo credits; acknowledgments; appendix; index; paperback.

¶ *One of the most popular and widely read of all Colorado River books is "Through The Grand Canyon From Wyoming To Mexico" [Farquhar #53]. It was written by Ellsworth Kolb and first published in 1914 with twenty-eight reprints since. With the goal of taking both still and moving pictures, Ellsworth and his brother Emery started from Green River, Wyoming and arrived at the foot of the Bright Angel Trail in the middle of winter. They both kept daily journals of the complete trip. After about a month they continued the trip down river and this time Hubert R. Lauzon was with them to help with portaging and photography. Lauzon also kept a journal. Editor Suran has very carefully transcribed all three journals and included many letters sent by the boaters both before and during their long trip down the Green and Colorado Rivers. This was a "learn as you go project" because all the expedition's members were doing their first white water*

44

*boating; their lack of experience caused many very exciting
moments as they came down the rivers. The Ellsworth Kolb
book is a great read but the journals are even better. Two
hundred copies were in hardback, numbered, signed and in
slipcase with "Every Rapid Speaks Plainly," #52 in this list.
These two books are the first volumes of "The Colorado River
Chronicles" series.*

55. LAVENDER

RIVER RUNNERS OF THE GRAND CANYON. By David
Lavender. Tucson: University of Arizona Press, [cloth w/
dust jacket], and Grand Canyon: Grand Canyon Natural
History Association, [paperback]. 1985. Pp. 133; ills.; map;
epilogue; acknowledgments; bibliography; index.

¶ *They are all here: from the journey that James White may
have made and the journeys John Wesley Powell and his
crews did make, to the speed runs made by modern boatmen
on the 1980 and 1983 high water. Read the stories of Stanton,
Flavell, "Hum" Woolley, Stone, Russell and Monett, the
Kolbs, and the many that followed for almost as many reasons.
Each chapter is full of exciting experiences told by a writer
very familiar with the subject matter: the Colorado River and
the country through which it runs. Many historic photographs
are included in this book which was a needed contribution to
the history of running the river through Grand Canyon.*

56. LEYDET

TIME AND THE RIVER FLOWING: GRAND CANYON. By
François Leydet, edited and with a foreword by David
Brower. San Francisco: The Sierra Club. 1964. Pp. 176;
ills.; endpaper maps; epilogue; appendixes.

¶ *One of the Sierra Club Exhibit Format Series, this volume was designed to draw attention to the Bureau of Reclamation's plans to build two dams in Grand Canyon. François Leydet has written a good summary about the Canyon's history, the National Park, and the goals of the Bureau of Reclamation. Included is a three-week float trip in dories that just about ran out of water. While his group was on the river the gates at Glen Canyon Dam were closed enough that only 1000 c.f.s. of water was coming through. P. T. Reilly was on this trip. The photographs are by several well known photographers and are all excellent. Many pages carry quotes from such people as Clarence Dutton, Frank Waters, Joseph Wood Krutch, John Wesley Powell, John Van Dyke, Aldo Leopold, Loren Eiseley, Wallace Stegner, and others. The quotes will lead you to many good books and much good reading.*

57. MASLAND

THE GRAND CANYON. By Frank E. Masland, Jr. Privately printed by the author and passed out to his friends. Circa 1964. Pp. 54; white self-wraps.

¶ *The first two chapters are a total re-writing of Masland's 1948 trip with Norm Nevills that he wrote about in "By the Rim of Time" [Farquhar #57]. This version is equally enjoyable. Chapter three is a trip taken in 1954 with 14,000 c.f.s. of water, using aluminum boats powered by outboard motors. Among others in the party were "Dock" Marston, Bill Belknap, Rod Sanderson, and Willie Taylor. The last chapter is an exciting story of a 1956 trip with 75,000 c.f.s. At the end of the book in an addendum is a note about Willie Taylor who passed away on the 1956 trip and was buried near President Harding Rapid. Five Quail Books did a facsimile reprint of this title in 1997.*

58. MAURER

THE WILD COLORADO. THE TRUE ADVENTURE OF FRED DELLENBAUGH, AGE 17, ON THE SECOND POWELL EXPEDITION INTO THE GRAND CANYON. By Richard Maurer. New York: Crown Publishers, Inc. 1999. Pp. 120; ills.; maps; author's note; suggested reading; index; picture credits.

¶ *This title, written for young readers, is a condensed version of the story Dellenbaugh tells in his classic "A Canyon Voyage" [Farquhar #45]. The maps, illustrations and photographs, several by Hillers, are of the time period of the trip and a few will be difficult to find in other publications. As well as writing books for young readers the author is also a photo researcher.*

59. NELSON

ANY TIME ANY PLACE ANY RIVER. THE NEVILLS OF MEXICAN HAT. By Nancy Nelson. Foreword by Barry Goldwater. Flagstaff: Red Lake Books. 1991. Pp. ii, 83; ills., map; endnotes; index; paperback.

¶ *This is a good biography of the pioneer of commercial river running on the San Juan and the Colorado River through Glen Canyon and Grand Canyon. Nevills did this in boats that he designed and built. He also took the first women through Grand Canyon. Nevills took Barry Goldwater on his first trip through the Canyon; Dock Marston went first as a paying passenger and later became a boatman for Nevills. Remember, these trips were on the old Colorado River before the upstream dams controlled the flow and you still had to let the water set in the bucket a little while before you wanted to drink it.*

60. NIMS

THE PHOTOGRAPHER AND THE RIVER 1889-1890. THE
COLORADO CAÑON DIARY OF FRANKLIN A. NIMS WITH THE
BROWN-STANTON RAILROAD SURVEY EXPEDITION. Edited
by Dwight L. Smith. Santa Fe: Stagecoach Press. 1967. Pp.
75; ills.; map; notes.

❡ *Nims was the official photographer on both the ill-fated
first attempt to survey the Colorado Canyons, which left
from Grand Junction, Colorado, and the second attempt
which left from Crescent Creek in Glen Canyon. On the first
attempt Nims came down the Green from Green River, Utah
with Brown and the main party and met the survey at the
Confluence. No photographs were taken on the Grand above
the Confluence. Four of his photographs are in this little book.
Nims wrote diaries in short-hand so he could, in a fairly short
time and small space, tell quite a lot about the day's activities.
Editor Smith used typescript copies that Nims had made in
editing this version for publication. Nims used one of the
boatman's diaries to fill in the time period after his fall, injury,
and evacuation. One minor point about Nims that Smith
found in the Stanton papers seems of special interest and most
appropriate. "He is a prominent Odd Fellow, and holds both
life and accident insurance." Jack Rittenhouse has given us
his usual fine quality production in this very important piece
of river and canyon history. Six hundred copies were printed.*

61. SADLER

THERE'S THIS RIVER. GRAND CANYON BOATMAN STORIES.
Edited by Christa Sadler. Flagstaff: Red Lake Books.
1994. Pp. 183; ills.; maps; information about Grand
Canyon; acknowledgments.

48

¶ *These are short, entertaining and no doubt "totally true" stories of the River and the Canyon, written by some boatmen who are men and some boatmen who are women. These people bring a different perspective to the river experience than the paying passengers because of their many trips and their knowledge of the River and the Canyon. The stories they tell to the passengers increase understanding and enjoyment while on the river and afterward. They are very enjoyable stories whether you have been on a river trip or not.*

62. STANTON

Down the Colorado. By Robert Brewster Stanton. Edited and with an introduction by Dwight L. Smith. Norman: The University of Oklahoma Press. 1965. Pp. xxv, 237; ills.; maps; index.

¶ *Until this volume, the only part of Stanton's thousand-plus-page magnum opus titled "The River And The Canyon" that had been published was in "Colorado River Controversies," edited by James M. Chalfant [Farquhar #50]. That title, financed by Julius Stone, dealt only with the James White claim and with the affair at Separation Rapid on the Powell trip of 1869. In "Down The Colorado" Smith has given us "with a minimum of editorial aids" the first eleven chapters of the second volume of Stanton's monumental work. These chapters covered the explorations and survey for the water-level railroad that ran from Grand Junction down the Grand River to the Confluence, and then down the Colorado all the way through Grand Canyon. On the original survey Brown, Stanton, and the rest of their group came down the Green River from Green River, Utah to the Confluence. [Frank C. Kendrick ran the survey from Grand Junction to the Confluence. See the article by Helen J. Stiles in "Colorado Magazine," Vol. xli, No. 3, Summer, 1964 for the Kendrick diary.] Here then, with information from Stanton's personal knowledge of the river, his notes, engineering data, and photographs, is the story he wrote about the survey expedition down the river. The unpublished portions of the two-volume manuscript covered earlier and later river history and both periods have now been written about by later writers. Therefore it is probable that the problem Stanton had in finding a publisher for the complete work will continue. This volume is one of the more important primary sources for the history of the Colorado River.*

63. STANTON

THE COLORADO RIVER SURVEY. ROBERT B. STANTON AND THE DENVER, COLORADO CANYON & PACIFIC RAILROAD. By Robert Brewster Stanton. Edited by Dwight L. Smith and C. Gregory Crampton. Salt Lake City: Howe Brothers. 1987. Pp. xiv, 305; ills.; maps; bibliography; index.

❡ *These are the field notes made by Stanton during the survey. They are very interesting and were one of the sources he used in writing his still unpublished manuscript, "The River and the Canyon." Stanton was a very faithful and detailed writer of his daily notes. The notes are from the ill-fated first survey attempt, where Railroad President Brown and two crew members drowned, and the second successful attempt with Stanton in charge. Early in the second trip photographer Nims was injured in a fall and had to be taken out. Stanton took over the duties of photographer and his first exposure is printed on page 123. The editors have done an extensive and very successful search of all the possible sources of information, including all of Stanton's known writings. From this information they have included footnotes, where needed, which explain what the day-by-day notes are about and why they are important. A list of Stanton's published writings is included in the bibliography. One hundred and twenty-five copies of this title were bound in leather and signed by the editors.*

64. STAVELEY

BROKEN WATERS SING. REDISCOVERING TWO GREAT RIVERS OF THE WEST. By Gaylord Staveley. Boston: Little, Brown and Company, A Sports Illustrated Book. 1971. Pp. xvii, 238; ills.; map; index.

¶ *In the centennial year of Powell's first voyage Gaylord Staveley led a group down the Green and Colorado as a celebration. At the time Staveley was the son-in-law of the late Norman Nevills and he used some of the Nevills boats or boats of similar design for this trip. They put in just below Flaming Gorge Dam on June 24 and, after portaging around the Powell reservoir, ran to Diamond Creek in Grand Canyon, spending forty-three days on the river. His descriptions of running the rapids are excellent. Especially appreciated are those on the upper river which few river runners since Powell have described in any detail. Staveley had floated Glen Canyon and Grand Canyon before the dam and tells what riding "waste water," as he calls the water that comes through the dam, is like compared to the natural river. It seems the rapids in both rivers look different from the rowing compartment of a Nevills boat compared to the view from the back of a thirty-foot raft with your hand on the tiller of an outboard motor.*

65. STEGNER

BEYOND THE HUNDREDTH MERIDIAN. JOHN WESLEY POWELL AND THE SECOND OPENING OF THE WEST. By Wallace Stegner with an introduction by Bernard DeVoto. Boston: Houghton Mifflin Co., The Riverside Press. 1954. Pp. xxiii, 438; ills.; maps; notes; index.

¶ *In this, the second major biography of Powell [see Farquhar #46], Stegner gives full coverage to the river trips – both the 1869 exploration and the 1871-72 survey with* its later work done on the Plateau. Stegner's book came at a time when those favoring more dams and those favoring wild rivers were becoming aware of the battles ahead. This biography helped to polarize both points of view. Stegner points out the significance of Powell's thinking and writing on the problems of settlement of the arid lands of the West. The problem of overcoming the denial of reality by people like William Gilpin, Captain Samuel Adams, and several members of Congress is an important part of the book. Stegner highlights the political maneuvering that was necessary for Powell to move his ideas forward when he was working as head of both the Geological Survey and the Bureau of Ethnology. If you want to know John Wesley Powell, read this book. In 1982 a signed limited edition was issued in slipcase.

66. SUMNER

Traveler in the Wilderness. By Cid Ricketts Sumner. New York: Harper and Brothers, Publishers. 1957. Pp. 248; endpaper maps; map.

❡ *What do you do for a change of pace if it is the mid-1950s, you are a woman, you are sixty-four years of age, and you have never been west of Chicago? Why, you sign up to go down the Green and Colorado Rivers with the Eggert-Hatch River Expedition. The plan was to take pictures of the canyons from Green River, Wyoming through Grand Canyon before the dam builders did any more of their work. Unfortunately the expedition had to stop at Lee's Ferry due to low water. Charles Eggert was shooting motion picture film but he and others also did many stills. Charles Eggert wrote a little article about this trip that was published in the January-*

March 1956 "National Parks Magazine." A slightly longer article with several Philip Hyde photographs was published in the November 1958 "Sierra Club Bulletin." The Sierra article is also printed in "The Glen Canyon Reader" edited by Mathew Barrett Gross and published by the University of Arizona Press, 2003. Charles Eggert has written his complete story and has a manuscript that as yet has not been published. Consequently, Mrs. Sumner's book is the only major work that has been published about this important trip. She writes with skill and was a very good observer of experiences totally new to her. She brings many insights into river running and the Canyon Country in this record of a little known expedition. In 1958 a British edition with identical text was published in London by Macdonald and Co.

67. TEAL

BREAKING INTO THE CURRENT. BOATWOMEN OF THE GRAND CANYON. By Louise Teal. Tucson: The University of Arizona Press. 1994. Pp. xvi, 178; ills.; bibliography; acknowledgments.

❡ *Georgie White was first and for a long time the only woman leading trips down the Colorado River through Grand Canyon. At the time Teal wrote this book, fifteen per cent of the people guiding on commercial trips in dories or rafts through Grand Canyon were women. The stories Teal presents are from eleven of the first boatwomen and they share with you why they came and why they stayed. Many other boatwomen were interviewed and their experiences are included in the full story. For many it was not an easy time. This was and had always been a male dominated profession and many wanted it to stay that way. As it turns out, brains are equal to or even more important than brawn.*

68. WATSON

THE PROFESSOR GOES WEST. REPORTS OF MAJOR JOHN WESLEY POWELL'S EXPLORATIONS: 1867-1874. Compiled and edited by Elmo Scott Watson with a foreword by Julia S. Watson. Illinois Wesleyan University Press. 1954. Pp. 138; appendix.

¶ *The reports of 1867-68 are of the student field trips led by Major Powell to Colorado. The 1869 reports, of course, involve the river trip and the ones following are concerned with exploration and survey work north of Grand Canyon. The messages, many not written by Powell, are in the form of letters about the activities of the Major and the people with him, written to and published by newspapers or Illinois Wesleyan University publications. Some of this is previously unpublished material. There is information of interest here that will not be found in this much detail, if found at all, in other Powell biographies.*

Of special interest is the following statement from an interview soon after the completion of the 1869 trip and first published in the "Chicago Republican": "Major Powell discredits the narrative of the passage of the Colorado published a short time ago in 'Lippincott's Magazine', descriptive of the trip of one James White, who alleges he went through the great canyon." That issue of "Lippincott's Magazine" was published in December 1868, about six months before Powell started on his first river journey. From their journals it seems some of Powell's men knew of White's story. Comparing the publishing date of the magazine and the departure date of Powell's first trip one would wonder if Powell might have seen the magazine. Did he know that he could probably get down the river but that he wouldn't be first, or did he brush the White claim off as a hoax?

In 1950 Illinois Wesleyan University Press published the title "The Illinois Wesleyan Story 1850-1950" also authored by Elmo Scott Watson. One chapter is titled "The Professor Goes West" and is an abbreviated version of the story.

69. WEBB

IF WE HAD A BOAT. GREEN RIVER EXPLORERS, ADVENTURERS, AND RUNNERS. By Roy Webb. Salt Lake City: University of Utah Press, Bonneville Books. 1986. Pp. xi, 194; ills.; acknowledgments; introduction; notes; bibliography; index; paperback.

¶ *Webb relates the history of European contact with the Upper Green in the first part of this book. Following that are all the river runners from Ashley, Manly, and Powell, to Bus Hatch and Norm Nevills's pioneer trips. The surveyors and dam builders are included along with the fight to save Dinosaur National Monument. There are people and events in this book that many river history buffs may not have heard of, but who have exciting stories to tell. The last chapter titled "The Other 'Place No One Knew'" is of special interest. It relates the lack of attention and concern in regard to the building of Flaming Gorge Dam and the drowning of Flaming Gorge, Horseshoe, Kingfisher, and Red Canyons. Many quotes from river runners who recorded their descriptions of these canyons in journals and diaries complete the story.*

70. WEBB

RIVERMAN. THE STORY OF BUS HATCH. By Roy Webb. Rock Springs: Labyrinth Publishing. 1989. Pp. x, 158; ills.; maps; notes on sources; footnotes; photo credits; paperback.

¶ *Bus Hatch was one of the first of the group we now call "river runners" who wanted to know what was around the next bend of the river. For Hatch, running the river got started as a way to get to a good fishing hole. He and some friends first used logs to carry them down to a particular spot and they would then walk back to where they had started. Bus*

Hatch and his cousin Frank Swain knew Parley Galloway and, after less help from Galloway than they would have liked, built their first boat. In August of 1931, with Bus Hatch as leader, Frank Swain, Tom Hatch, and Cap Mowrey put the new boat in just below Flaming Gorge. Early in the four day trip they upset, lost all their food and fishing gear, and were getting very hungry when they reached the mouth of Split Mountain Canyon. They had learned many lessons. The group had the desire to go again and eventually hoped to go all the way through Grand Canyon. It took several years to make that dream come true and this biography has the complete story. A second edition was published in 1990 in a hard back, numbered, and signed printing of fifty copies.

71. WELCH/CONLEY/DIMOCK

THE DOING OF THE THING. THE BRIEF BRILLIANT WHITEWATER CAREER OF BUZZ HOLMSTROM. By Vince Welch, Cort Conley, and Brad Dimock. Flagstaff: Fretwater Press. 1998. Pp. xi, 290; ills.; end paper maps; acknowledgments; index.

❡ *After reading this exceptional biography it is difficult to believe that three different authors contributed to it. It flows as if from one mind and one pen. It is a remarkable story of a remarkable young man. Haldane "Buzz" Holmstrom was the first – and there haven't been many since – to float the Green and Colorado Rivers solo. He picked out the log, had it sawed into lumber, then designed and built his own boat. In the portions of his river journals included in the text he tells of his anxiety, his awe, and the beauty of the canyon on his solo trip of 1937. The 1938 trip with Amos Burg, where Holmstrom ran Lava Falls, is detailed as well. In his short life Holmstrom had many other river experiences and all are included.*

72. WESTWOOD

WOMAN OF THE RIVER. GEORGIE WHITE CLARK—WHITE WATER PIONEER. By Richard E. Westwood with a foreword by Roy Webb. Logan: Utah State University Press. 1997. Pp. xiv, 304; ills.; maps; appendix; notes; related works; index; paperback.

¶ *If you had talked to other "river people" when Georgie was running the river, it would have been difficult to find someone that did not have a well formed opinion about her. That opinion may have been positive or it may have been negative but not many were neutral in their feelings. Dick Westwood has written a biography that is well balanced and tells both sides of Georgie. Regardless of anyone's feelings about the "Woman of the River" she is a part of the history of running the rivers and this book puts her in proper prospective.*

73. WHITE

a

HELL OR HIGH WATER. JAMES WHITE'S DISPUTED PASSAGE THROUGH GRAND CANYON 1867. By Eilean Adams. Logan: Utah State University Press. 2001. Pp. 221; ills.; maps; appendix; chapter notes; sources; author's note.

b

AMERICAN WEST MAGAZINE, Vol. xix, No. 6, November/ December 1982, "FIRST THROUGH THE GRAND CANYON." By David Lavender.

c

THE BULLETIN, Missouri Historical Society, Vol. 17, No. 4, July 1961, "FIRST MAN THROUGH THE GRAND CANYON." By Dr. Harold A. Bulger. [There is also an off-print of this article.]

d

THE GRAND CANYON. AN ARTICLE GIVING THE CREDIT OF FIRST TRAVERSING THE GRAND CANYON OF THE COLORADO TO JAMES WHITE, A COLORADO GOLD PROSPECTOR, WHO IT IS CLAIMED MADE THE VOYAGE TWO YEARS PREVIOUS TO THE EXPEDITION UNDER THE DIRECTION OF MAJ. J. W. POWELL IN 1869. By Thomas F. Dawson. Phoenix: Five Quail Books. September, 2001. Pp. 67 plus seven page bibliography, paperback; and Prescott: Five Quail Books. November, 2002. Pp. 67 plus nine page bibliography and a one page addendum and with the report of January 6, 1868, to J. D. Perry, Esq. from C. C. Parry as printed by the St. Louis Academy of Natural Science tipped in. Thirty copies in boards, signed by Dan Cassidy of Five Quail Books, five of which are in slipcase.

e

FIRST THROUGH THE GRAND CANYON. By Richard E. Lingenfelter with a foreword by Otis Marston. Los Angeles: Glen Dawson, Early California Travel Series XLV. 1958. Pp. 119; ills.; bibliographical notes; index.

f

"THE RELUCTANT CANDIDATE – JAMES WHITE, FIRST THROUGH THE GRAND CANYON." by O. Dock Marston. BRAND BOOK NUMBER THREE THE SAN DIEGO CORRAL OF THE WESTERNERS. San Diego: The San Diego Corral Of The Westerners. 1973. Pp. 166-176. ills.

g

FIRST TO JOURNEY THROUGH THE GRAND CANYON: THE LIFE STORY OF JAMES WHITE. By Brad Smith. Cochise: Brad Smith. 2001. Pp. 28; ills.; maps; bibliography; paperback.

¶ *Did he or didn't he? This has been the question since 1867. The only accepted facts are that a scratched, bruised, sunburned, and nearly naked James White was pulled from* the Colorado River at Callville, Nevada the afternoon of September 7, 1867. White talked to or was interviewed by several people in the next few weeks. His basic story was that he, Captain Baker, and George Strole had left Baker Park (now Silverton) in southwestern Colorado in late July. The three prospectors traveled and prospected down to the Mancos River and then to the San Juan. When the San Juan canyoned they climbed out and went overland, attempting to reach Grand River. At some point the trio was attacked by Indians and Baker was killed. Leaving their horses but taking a few provisions, White and Strole worked their way down to a river. Here they made a raft by tying logs together with ropes, and embarked on the stream. According to White, Strole was washed off the raft and drowned early in the journey. After a period of time which seems uncertain, White arrived by river at Callville.

The author of a is one of James White's granddaughters. She has consulted every known source in her research and combined it with information that came to her as family history. A very important source is research by the late Robert C. Euler done from 1972-1975 and published here for the first time. Adams has presented a very logical discussion and if this material was being judged as a debate she would certainly receive very high marks. In b Lavender gives a well balanced presentation of both the early reports and the later criticisms of White's supposed journey. A slightly different form of this story appears in Lavender's book, "Colorado River Country," also published in 1982. In c Dr. Bulger follows White's letters and early interviews better than most of his biographers but like Lingenfeltger, the author of e, he is not quite as familiar with the geography of the region as might be desired. In e Lingenfelter gives a well documented presentation of what White might have done on a river journey and presents the most complete story written until a. In f Marston has collected information from all the previous writers and points out some errors in interpretation of information and in some cases the publishing of incorrect information. In g Smith relies heavily on *Lingenfelter's work but does have some new information on Captain Charles Baker. Yes,* d *is also Farquhar #39, but the Five Quail reprint includes an extensive bibliography of publications by other authors who have mentioned James White in their writings. It is an interesting list and of value.*

74. WORSTER

A RIVER RUNNING WEST. THE LIFE OF JOHN WESLEY POWELL. By Donald Worster. New York: Oxford University Press. 2001. Pp. xiii, 673; ills.; notes; bibliography; index.

¶ *This is the most detailed and complete of all the biographies of Powell. It brings out his best qualities and his faults. His early life and the influence his strongly religious parents' philosophy had on him, his eagerness to seek scientific knowledge, his military career, and his explorations are each given the importance they deserve. Much of the story involves his years in Washington. Here he was fighting battles that in many ways involved greater difficulties than the ones he fought in the Civil War or the struggle of going down the rivers. Everything you would wish to know about him is in this honest, thorough, and well balanced look at one of the truly giant figures of the late nineteenth century.*

Part V

THE CANYONS APPRECIATED
AND UNDERSTOOD

75. AITCHISON

A WILDERNESS CALLED GRAND CANYON. By Stuart
Aitchison with a foreword by Jim Ruch. Photography by
Dick Dietrich and others. Stillwater: Voyageur Press, Inc.
1991. Pp. 127; map; ills; further reading.

¶ *This is a natural history book and if text books were written
like this a larger number of students would be interested
in the subject. Aitchison is a naturalist and photographer
who has explored Grand Canyon and the Colorado Plateau
for over twenty-five years. He brings an understanding of
the relationships of all life in and around the canyon and
combines that with the writing skills to inform both the
professional and the layman. The color images included are
well matched to the subject matter presented in the various
chapters. This was a needed addition to the literature of the
canyon country.*

76. ATON/MCPHERSON

RIVER FLOWING FROM THE SUNRISE. AN ENVIRONMENTAL
HISTORY OF THE LOWER SAN JUAN. By James M. Aton
and Robert S. McPherson. Foreword by Donald Worster.
Logan: Utah State University Press, 2000. Pp. 216, map;
ills; notes; bibliography; index; paperback.

¶ Man has lived along the lower San Juan River for at least 12,000 years. The Navajo, historically new to the southwest, were the first to cause serious environmental damage. They acquired sheep from the Spanish settlements and overgrazed much of the already sparse vegetation of the area. When the whites came in with their cattle herds the Navajo almost looked like good guys. Add to those things the miners, oil drillers, and dam builders, and a more complete picture is drawn of the environmental abuse along the river. Although this book is not the first writing to tie history and the environment together, this is one of the first that actually set that as its goal. It points out, very clearly, that what man does to the environment is related directly to what the environment can do for man. This is a simple and basic concept but one our culture has been very slow in understanding. Greed is a terrible thing! The authors have presented a full and well researched history of one of the least known major tributaries of the Colorado River. It is a total history, not only of the people who have lived along it, but also a history of the river itself. The people changed the river and the river changed the people. For too long these two entities worked against each other. The authors point out that each will be far better off if they work together. Environmental history is an idea whose time has come. Aton and McPherson have set a very high standard for future writers to follow.

77. BRIAN

RIVER TO RIM. A GUIDE TO PLACE NAMES ALONG THE COLORADO RIVER IN GRAND CANYON FROM LAKE POWELL TO LAKE MEAD. By Nancy Brian. Flagstaff: Earthquest Press. 1992. Pp. iii, 176; ills; appendixes; bibliography; index; paperback.

¶ What's in a name? What is that called? Why is it called that? These questions and many others are answered in this book. *More than 600 place names are listed in order by river mile and located on river right or river left. Brian gives the history of the name as well as any interesting happenings that took place at that location. This title was chosen over the book "Arizona Place Names," by Will Barnes, revised and enlarged by Byrd H. Granger because it contains many additional Grand Canyon related entries and the smaller format is easier to use. Because different information is sometimes given, the two would make excellent shelf mates.*

78. BUFF

THE COLORADO: RIVER OF MYSTERY. By Mary and Conrad Buff. Los Angeles: The Ward Richie Press. 1968. Pp. 86; ills.; map.

¶ *Here is a well researched and well written account of both the natural and human history of the Colorado River and Grand Canyon for younger readers. The many illustrations are very nice and the map is quite adequate. The total package is what we have come to expect from The Ward Richie Press.*

79. BUTCHART

GRAND CANYON TREKS. 12,000 MILES THROUGH THE GRAND CANYON. By Harvey Butchart, edited and designed by Wynne Benti, with a foreword by Harvey Butchart and an introduction by Wynne Benti. Bishop: Spotted Dog Press. 1998. Pp. 288; ills.; maps; appendixes; references; index; paperback.

¶ *In 1970, 1975, and 1984, La Siesta Press of Glendale, California issued the titles "Grand Canyon Treks," "Grand Canyon Treks II," and "Grand Canyon Treks III." The present volume is a re-publication of the material from these three slender little volumes with some added material, some updating, and a few different photographs. The late Dr. Butchart is the acknowledged dean of all Grand Canyon backcountry hikers and backpackers. This is not a guide book as most guide books go. It could be helpful in planning backcountry routes, but beware: Butchart is a master of brevity and understatement in route descriptions. Rather, it is a book of experiences and observations below the rim with some interesting history thrown in. He has marked all of his routes on maps [now housed at Cline Library, Northern Arizona University] so that his knowledge can be shared with others. Some of his treks were with companions but many were made alone, possibly because he could find no one that wanted to accompany him over the difficult route to his planned destination. Given Butchart's almost 50 years and over 12,000 miles of hiking in the Canyon, it would be difficult – though perhaps not impossible – to go somewhere the author has not been. Someday, someone may break his record; someday, maybe. A second printing was done in 2000 with a few minor corrections.*

80. CAROTHERS/BROWN

THE COLORADO RIVER THROUGH GRAND CANYON. NATURAL HISTORY AND HUMAN CHANGE. By Steven W. Carothers and Bryan T. Brown with a foreword by Bruce Babbitt. Tucson: The University of Arizona Press. 1991. Pp. xix, 235. map; ills; appendixes; notes; index.

❡ *Carothers and Brown have written a very thorough study of the impact that man in general, and the Glen Canyon Dam in particular, have had on Grand Canyon's native flora and fauna and on the experiences of recreational users of the river. This is a landmark study and will be of great value in the future for comparing conditions that exist twenty, fifty, one hundred, or more years from now. Some of the suggestions resulting from this study are already being put into action.*

81. CHILDS

STONE DESERT. A NATURALIST'S EXPLORATION OF CANYONLANDS NATIONAL PARK. By Craig Leland Childs. Englewood: Westcliffe Publishers. 1995. Pp. 198. ills; selected bibliography; paperback.

¶ *If you have never been to Canyonlands National Park, read this book before you go. You will see more while you are there even if you don't spend much time (or maybe not any) in the backcountry. If you have been to the park, read this book and realize what you missed that you can see and do on your next visit. Childs understands the relationships between all the elements: the rocks, the plants, the animals, the wind, the cold and the heat, the water and the lack of water, time, and space. Canyonlands has some of the oldest and most remarkable prehistoric rock art on the continent and Childs visits many of the sites. He explains how the people who lived here a thousand years ago were able to wrest a living from this harsh environment. This book will fit in your backpack.*

82. CHILDS

THE SECRET KNOWLEDGE OF WATER. DISCOVERING THE ESSENCE OF THE AMERICAN DESERT. By Craig Childs. Seattle: Sasquatch Books. 2000. Pp. xvi, 288. ills.; bibliography; index.

¶ *A group of truly fascinating experiences, most of them on the Colorado Plateau. Childs is a naturalist who has great powers of observation, the knowledge to understand what he is seeing, and the writing style and skill to share it with others. He puts himself into situations that are often unique and sometimes dangerous. But with his knowledge of the workings of water in the desert and canyons, Childs shares his experiences and you almost feel as if you are there. One selection titled "Water that Waits" is one of the most thought provoking pieces of writing in all the literature of the Colorado Plateau. The setting is a place John Wesley Powell named "The Thousand Wells" but things are going on there that Powell could never have dreamed about. Still Powell, with his scientific mind, would certainly have been interested. [If you wish to read more on this topic look in the winter 1999-2000 "Plateau Journal" and read the article by Dr. Tim Graham.] Water has been of primary importance in shaping the Colorado Plateau's many features and this book helps us understand the role it has played.*

83. CHILDS/LADD

GRAND CANYON. TIME BELOW THE RIM. Text by Craig Childs with photographs by Gary Ladd. Phoenix: Arizona Highways Books. 1999. Pp. 192. maps; ills.; index of photographs and illustrations.

¶ *Childs and Ladd have spent a great deal of time exploring routes into Grand Canyon backcountry, where there are few if any trails, to gather the materials that make up this book. Childs with his notebooks and Ladd with his cameras were not physically traveling together, but their thoughts were in harmony. The text and the photographs take us places most of us will never go, but to see them through the pages of this large format book is a very rewarding and most enjoyable experience.*

84. COLLIER

Water, Earth, and Sky. The Colorado River Basin. Photographs by Michael Collier with a foreword by David L. Wegner and essays by Michael Collier, John C. Schmidt, E. D. Andrews, Richard A. Valdez, Lawrence E. Stevens and Ellen Meloy. Salt Lake City: The University of Utah Press. 1999. Pp. 126. map; ills.; selected readings; contributors.

¶ *For this book Michael Collier flew his forty-year old Cessna airplane on many trips from the fountain peaks of Wyoming and Colorado to the Gulf of California. After doing so he selected this magnificent collection from about twelve thousand aerial photographs of the Colorado River basin. These views from the air provide a totally different perspective; by viewing them as a collection you can soon see how the whole basin is connected and shaped by water. But this is not just a collection of photographs. The essays by experts in their fields would have made an excellent book on its own merits. Combining the essays with the photographs has made this title one of the outstanding books about the Colorado River country.*

85. CRAMPTON

Land of Living Rock. The Grand Canyon and the High Plateaus: Arizona, Utah, Nevada. By C. Gregory Crampton. New York: Alfred A. Knopf. 1972. Pp. xi, 267, vii. ills.; maps; notes; bibliography; index.

¶ *This volume and the one following are soul mates and should be shelf mates as well. In this volume Crampton has covered the geological history and the human history of Grand Canyon and the high plateaus. He includes enough information to inform you but not so much that your interest is lost. In fact, you may end each chapter wanting*

*supplementary knowledge and the fine bibliography will
certainly guide you there. Crampton takes you to several
places very few have heard of and fewer have been. Many of
these places are included in the excellent photographs that
are in the book. There are few if any people that have seen
more of the Colorado Plateau than Gregory Crampton. He
understood and appreciated what he saw.*

86. CRAMPTON

a

STANDING UP COUNTRY. THE CANYON LANDS OF UTAH AND
ARIZONA. By C. Gregory Crampton. New York: Alfred A.
Knopf and The University of Utah Press in association
with the Amon Carter Museum of Western Art. 1964.
Pp. xv, 191, iv; ills.; maps; notes; bibliography; index.

b

STANDING UP COUNTRY. THE CANYON LANDS OF UTAH AND
ARIZONA. By C. Gregory Crampton. Tucson: Rio Nuevo
Publishers. Second edition, 2000. Pp. viii, 118; ills.; maps;
suggested reading; illustration credits; index; paperback.

❡ *If one book had to be chosen above all the others to best
describe and interpret the canyon country of southeastern
Utah and northeastern Arizona, no better choice could be
made than this book. Crampton knew the area well. He covers
the prehistoric people and the present Native Americans and
includes the Spanish entradas, government surveys, miners,
settlers and, as he called it, the "Symphony in Sandstone."
The first edition, a, is well illustrated using both black-and-
white and color photographs, many of the scenes being well
off the beaten path. A smaller format was used in b along with
all different photographs, most of them in color, and the text
has been edited from the first edition. In either edition this is a
book you will go back to again and again. This title and #85
were issued in a slipcased set with the dates 1983 and 1985.*

68

87. FLEISCHNER

SINGING STONE. A NATURAL HISTORY OF THE ESCALANTE CANYONS. By Thomas Lowe Fleischner. Salt Lake City: The University of Utah Press. 1999. Pp. xix, 212. ills.; maps; taxonomy of species mentioned in text; notes; references; index.

¶ *Natural History, very simply, is the history of all things natural. Fleischner visited the Escalante Canyons for fifteen years before the Grand Staircase – Escalante National Monument was formed. In this presentation he acquaints us with the intricate web that ties life and its surroundings together. His understanding and ability to convey these ideas could have come only after much observation and contemplation in the canyons. You will look at rocks, trees, desert streams, flowers, and butterflies differently after you read this delightful book. You will also know why, and be glad, the Escalante was protected.*

88. FLETCHER

THE MAN WHO WALKED THROUGH TIME. With photographs taken en-route by the author. By Colin Fletcher. New York: Alfred A. Knopf. 1967. Pp. 239. ills.; map; appendix.

¶ *It is approximately eighty-five miles from Hualpai Hilltop to Point Imperial the way the raven might go, but Fletcher walked below the rim and the route turned out to be a little farther. The two-month journey exposed him to the Canyon's quickly changing weather and its difficult terrain. He experienced the lack of water along much of the route, and the wildlife that few people see. The journey also revealed the Canyon to Fletcher for he viewed it alone, from vantage points rarely visited, with the time to contemplate the significance of what he was seeing. Although he didn't spend a year he probably would agree with Powell that "It is a region more difficult to traverse than the Alps or the Himalayas, but if strength and courage are sufficient for the task, by a year's toil a concept of sublimity can be obtained never again to be equaled on the hither side of Paradise."*

89. FOX

a

BELOW THE RIM. ONE WOMAN'S ADVENTURES IN THE GRAND CANYON. By Dottie Fox. Old Snowmass: Dottie Fox. nd. [circa 1990]. Pp. 98; paperback.

b

BELOW THE RIM. FOOTSTEPS THROUGH THE GRAND CANYON. By Dottie Fox. Basalt: Who Press. 2000. Pp. 159; ills.; maps; paperback.

¶ *These little books are not meant to be trail guides but they are excellent guides of a different order. They will not be of much value in helping you find the trail but they are excellent guides for helping you see and feel and appreciate all the things you can experience below the rim. Fox was an experienced backpacker when she first went below the rim, but was of an age when most backpackers start looking for trails that have shorter hills that aren't as steep. She is a keen observer of the things around her and has the ability in her writing to take the reader along but with a much lighter pack. There are a few added hikes in* b, *plus maps and illustrations by the author. It will also be much easier to find* b.

90. FROST

MY CANYONLANDS. I HAD THE FREEDOM OF IT. By Kent Frost. New York: Abelard-Schuman. 1971. Pp. 160; ills.; maps; end paper maps.

¶ *Kent Frost did a lot of hiking, exploring, and even some river running while he was growing up in southeastern Utah in the 1920s and '30s. These experiences developed into a career for him and his wife and he was taking people on tours into what is now Canyonlands National Park before the area was set aside. He was one of the people that worked at getting legislation passed for the park. He recalls stories of many of the local people he grew up around and gives you a real feel for the times. His stories of exploring and guiding in the canyon country are the next best thing to being there. This is a first-hand account of the early days in and around what was to become one of the great national parks on the Colorado Plateau. It is a hard book to put down. In 1997 there was a paperback printing with the same text by Canyon Country Publications, which includes seven historic photographs plus a recent photograph of the author.*

91. GHIGLIERI/MYERS

OVER THE EDGE: DEATH IN GRAND CANYON. GRIPPING ACCOUNTS OF ALL KNOWN FATAL MISHAPS IN THE MOST FAMOUS OF THE WORLD'S SEVEN NATURAL WONDERS. By Michael P. Ghiglieri and Thomas M. Myers with a foreword by Ken Phillips. Flagstaff: Puma Press. 2001. Pp. xiv, 408; ills.; maps; references; index.

¶ *This is a book that everyone who has hiked the trails or floated the river – or ever intends to do so – might wish to read or use as a reference. It is very enlightening reading. No one is invincible and second chances don't always come. This book is not designed to scare anyone from experiencing the Canyon but it will encourage everyone to have a very healthy respect for it. After all, many good lessons may be learned vicariously. As the Boy Scout motto tells us, "Be Prepared." You might also wish to consult "Fateful Journey. Injury and Death on Colorado River Trips in Grand Canyon" by Thomas Myers, Christopher Becker, and Lawrence Stevens.*

92. HOFFMAN

a

ARCHES NATIONAL PARK. AN ILLUSTRATED GUIDE AND HISTORY. By John F. Hoffman with a foreword by David D. May, photography by Frank L. Mendonca, illustrations by John D. Dawson. San Diego: Western Recreational Publications. 1981. Pp. vii, 104; ills.; maps; suggested readings and maps; paperback.

b

ARCHES NATIONAL PARK. AN ILLUSTRATED GUIDE. By John F. Hoffman with a foreword by David D. May, photography by Frank L. Mendonca, drawings by John D. Dawson. San Diego: Western Recreational Publications. 1985. Second edition, revised and enlarged. Pp. vii, 128; ills.; maps; appendix; suggested readings and maps; index; paperback.

¶ *Hoffman has included enough information on the geography and geology of Arches to give an adequate background for most visitors. Also included are chapters on the flora and fauna, prehistoric and historic Native Americans, Spanish entradas, exploration by fur trappers, and later by the Mormons. All these activities eventually evolved into area settlement. The history of the area that became the park is very well done. The drawings and photographs complement the text quite nicely and the maps and guide information will be found to be very helpful. Some new information has been added to* b *as well as several pages of color photographs not found in* a. *There was a hardback printing of* a *limited to 200 copies.*

93. HOUK

AN INTRODUCTION TO GRAND CANYON ECOLOGY. By Rose Houk with illustrations by Tony Brown. Grand Canyon: Grand Canyon Association. 1996. Pp. 56. ills.; maps; paperback.

¶ *This may be the first book which specifically addresses the ecology of Grand Canyon. Other offerings nibble around the edges but never get right into looking at the total web of life and the environment in which that web exists. There is enough information in this title to make you aware of the importance of having an understanding of ecology but not so much as to make you lose interest. As the title says, this is an introduction, but until someone puts together a complete study this little volume will stand as an important addition to the writings about Grand Canyon.*

94. HUTCHINSON

GRAND CANYON NATIONAL PARK. A PHOTOGRAPHIC NATURAL HISTORY. By Robert Hutchinson featuring the photography of fifteen leading photographers. San Francisco: Brown Trout Publishers. 1995. Pp. 127; ills.

¶ *This is not just another picture book, although the photographs are all outstanding. Hutchinson, a petrologist [one who studies the composition, structure, and origin of rocks], has divided the Canyon into six segments and has explained something about the rock formations in each. He explains how the different rock strata have helped determine the erosion patterns and therefore the shape of the Canyon, and the flora and the fauna that can exist in that particular type of environment. When hiking different trails in the Canyon or floating through on the river, the knowledge gained from this book will help you understand what you are seeing and why. The photographs are well matched to the information.*

95. KELSEY

a

RIVER GUIDE TO CANYONLANDS NATIONAL PARK AND VICINITY. FEATURING: HIKING, CAMPING, GEOLOGY, ARCHAEOLOGY AND STEAMBOATING, COWBOY, RANCHING & TRAIL BUILDING HISTORY. By Michael R. Kelsey. Provo: Kelsey Publishing. July 1991. Pp. 256; ills.; maps; paperback.

b

HIKING, BIKING AND EXPLORING CANYONLANDS NATIONAL PARK AND VICINITY. FEATURING: HIKING, BIKING, GEOLOGY, ARCHAEOLOGY, AND COWBOY, RANCHING & TRAIL BUILDING HISTORY. By Michael R. Kelsey. Provo: Kelsey Publishing. May 1992. Pp. 320: ills.; maps; paperback.

❡ *These two books are excellent as guides to help you get to many interesting places, but they are listed here for a different reason. Kelsey did extensive historical research including many oral interviews with people who helped make the human history. Before people were forced to attempt making a living in this canyon country, all the good farming and grazing lands were already taken. Farming or running livestock in this kind of country took tremendous effort and some special people to make that effort. Both of these books are as much involved with the human history of the Canyonlands country as they are guide books to help you find those places. If you are serious about looking for some of these places it might be well to have the USGS topographic maps as well as the maps in the books.*

96. KRUTCH

GRAND CANYON. TODAY AND ALL ITS YESTERDAYS. By Joseph Wood Krutch. New York: William Sloane Associates. 1958. Pp. 276; ills.; map.

¶ Krutch spent his professional career as a teacher at Columbia University and as a drama critic and writer. In 1950 he retired from those duties and moved to Arizona. He is a gifted observer, thinker and writer and in this book shares many of his thoughts related to the Canyon. The last chapter "What Men, What Needs" should be read by everyone, especially the country's decision makers.

97. LAMB

GRAND CANYON. THE VAULT OF HEAVEN. By Susan Lamb with photographs by Tom Bean, Gary Ladd, Larry Lindahl, and others. Grand Canyon: Grand Canyon Association. 1995. Pp. 67; photography credits; ills.; map; paperback.

¶ The text in this large format book has some geology, some natural history, some human history, and much food for thought and appreciation of the Canyon. The photographs are well selected to supplement the text. Every visitor should read this book before they come to the Canyon and again after they return home. It is a beautifully designed and produced book. A second printing was done in 1997 and bound in hardback.

98. LAMB

THE BEST OF GRAND CANYON NATURE NOTES 1926-1935. Edited by Susan Lamb. Grand Canyon: Grand Canyon Natural History Association. 1994. Pp. xvi, 167; ills.; suggested reading; paperback.

¶ In 1926 the Park Service staff began writing and publishing a small, few-page item called "Grand Canyon Nature Notes." This lasted for ten years. Original copies are very difficult to

find today as the number produced was not large and with paper covers they were very fragile. For this book Lamb has selected delightful vignettes from four basic categories: "Exploring Grand Canyon," "Earth Science," "Life Science," and "Human History." The writers were such people as E. T. Scoyen, Chief Park Ranger, Francois Matthes, U. S. Geological Survey, Superintendent M. R. Tillotson, Park Naturalists Edwin D. McKee and Glen Sturdevant, and many others. In the fall of 1994 Grand Canyon National Park, in cooperation with Grand Canyon Natural History Association, [later the Grand Canyon Association], again began publishing a modern version of "Nature Notes." It is included with the Association's newsletter, "Canyon Views."

99. LEE

Torrent in the Desert. By Weston and Jeanne Lee. Endpapers and maps by Don Perceval with a foreword by Edward B. Danson. Flagstaff: Northland Press. 1962. Pp. xvi, 186 plus the last plates; acknowledgments; ills.; maps.

¶ *The book is divided into roughly one-third text and two-thirds photographs. This is one of the few books that covers with photographs and text both the Green and Colorado Rivers from near their sources on down through the canyon country. These photographs were taken before the dams were built on the upper rivers or their tributaries, and Glen Canyon Dam was not yet complete. Therefore many of the pictures are of scenes no longer in existence. Most of the photographs are from the land but several are from the air or the river. The text describes many of the points of interest along the rivers and gives significant historical information as well. Some important side streams are also mentioned and photographed.*

100. MELOY

RAVEN'S EXILE. A SEASON ON THE GREEN RIVER. By Ellen Meloy. New York: Henry Holt and Company. 1994. Pp. 256. map; acknowledgments.

¶ *John Wesley Powell gave Desolation Canyon its name and it has stuck but Meloy finds much that doesn't fit under that title. She describes this great slash through the Tavaputs Plateau with much feeling and understanding of the things that are there. Meloy gives you information about the geography, archaeology, early settlement, local plant and animal life, and some timely comments on the politics of water in the West. The story moves both up and down the river from Desolation Canyon and at times away from the river. Meloy is one of the few visitors to Desolation Canyon since Powell to write about this area of the canyon country and in the book she will tell you about the raven.*

101. MUENCH/LAWRENCE

PLATEAU LIGHT. Photography by David Muench, text by James Lawrence. Portland: Graphic Arts Center Publishing. 1998. 116 plates and eight pages of text; photograph location and technical information.

¶ *There are only seven pictures of Grand Canyon in this folio-size volume but it is full of grand images from other areas. Many are in Canyonlands, Arches, Bryce Canyon, Capitol Reef, and Zion National Parks. Several are in the Grand Staircase/Escalante National Monument and Glen Canyon National Recreation Area and others somewhere on the Colorado Plateau. The text by James Lawrence is titled "Journey To A Sacred Place" and the words and images certainly take you there. Photographer Muench's goal is to capture "the timeless moment." The images in this book indicate he has been very successful.*

102. NASH

Grand Canyon of the Living Colorado. Edited by Roderick Nash with a foreword by David Brower. Photographs and journal by Ernest Braun with contributions by Colin Fletcher, Allen J. Malmquist, Roderick Nash, and Stewart Udall. With excerpts from the narration by David Brower, Jeffrey Ingram, and Martin Litton for the Sierra Club film "The Grand Canyon" accompanying the color plates. San Francisco: The Sierra Club. 1970. Pp. 143; ills.; map; Grand Canyon River Miles Lee's Ferry to Temple Bar.

¶ *This volume was part of the Sierra Club's effort to keep dams out of Grand Canyon. Allen Malmquist writes of exploring the backcountry of western Grand Canyon in the area of the proposed Bridge Canyon Dam site. Ernest Braun's journal and photographs cover an eighteen-day river trip with Martin Litton. Stewart Udall, then Secretary of the Interior, and Roderick Nash were on the same trip and contribute their impressions of the river and the Canyon. Nash gives a brief history of "The National Park Idea" and more specifically of Grand Canyon National Park, and discusses at some length the controversy of dams in Grand Canyon. Some of the Sierra Club ads opposing the dams are duplicated in the book. This item is part of an important record of the conservationists' battle to keep the Colorado free-flowing in the Canyon.*

103. NEUMANN

On the Rim. Looking for the Grand Canyon. By Mark Neumann. Minneapolis: University of Minnesota Press. 1999. Pp. xi, 373; ills.; notes; index.

¶ *Joseph Christmas Ives said of the Canyon that it "shall be forever unvisited and undisturbed." John Wesley Powell once referred to the Canyon as the "Great Unknown." Clarence Dutton wrote that " ... the Grand Canyon is the sublimest thing on earth." Theodore Roosevelt in his speech at the Canyon in 1903 said "What you can do is keep it for your children, your children's children, and for all who come after you, as one of the great sights which every American if he can travel at all should see." Quite a change in thinking in a period of a little less than fifty years. If you have been to Mather Point on a summer day in the last few years you might think everyone has taken T. R.'s advice, on that very day! Why do people come to Grand Canyon? That is the question that Mark Neumann has made a bold attempt to answer in this book. He delves into all the different facets in the human history of Grand Canyon and comes up with some logical conclusions, some of which may be rather surprising.*

104. PYNE

a

DUTTON'S POINT. AN INTELLECTUAL HISTORY OF THE GRAND CANYON. By Stephen J. Pyne. Grand Canyon: Grand Canyon Natural History Association. Monograph Number 5. 1982. Pp. v, 64. ills.; maps; notes; biographic note; index; paperback.

b

HOW THE CANYON BECAME GRAND. A Short History. By Stephen J. Pyne. New York: Viking Penguin, a member of Penguin Putnam, Inc. 1998. ills.; maps; afterword; appendix; notes; sources and further readings; index; illustration credits.

¶ *The first European contacts with Grand Canyon did not elicit comments of appreciation for its beauty or an understanding of its form. It took many years and visits by many people for that to happen. To quote from Dutton's "Tertiary History," [Farquhar #73] "Great innovations, whether in art or literature, in science or in nature, seldom take the world by storm. They must be understood before they can be estimated, and must be cultivated before they can be understood." Both these books outline the changes in attitude that have come about, and trace the significance of the ideas that have brought about those changes. In comparing the two books, you will notice many similarities. In b the author has expanded and re-written a, and several illustrations are the same in both books.*

105. RICHARDSON/CARRIER

The Colorado. A River at Risk. Photographs by Jim Richardson and text by Jim Carrier with a foreword by Wallace Stegner. Englewood: Westcliffe Publishers, Inc. 1992. Pp. 184. end paper maps; ills.

¶ *The old cliché, a picture is worth a thousand words, is very much up to date in this folio volume. Photographer Richardson drove, hiked, floated, and flew to get all the photographs used in this book and Jim Carrier's text certainly complements them. The photographer and writer cover the region from the headwaters of both the Green and Colorado all the way down to the Gulf of California. They find a working river, overworked in many instances, and a river on which millions of people come to play. They find a river dammed: a river siphoned off for irrigation, drinking water, and golf courses. They find a river that dies before it gets to salt water. This book is greatly expanded from an article in the June 1991 "National Geographic Magazine." Jim Carrier has another book, "Down the Colorado; Travels On A Western Waterway" published in 1989 that also tells a story of the river.*

106. SUTTON

THE WILDERNESS WORLD OF THE GRAND CANYON. "LEAVE
IT AS IT IS." By Ann and Myron Sutton with photographs
by Philip Hyde. Philadelphia: J. B. Lippincott Company.
1971. Pp. xii, 241; endpaper maps; acknowledgments;
appendixes; bibliography; index.

¶ *Writing over thirty years ago, these two naturalists were
aware of the special efforts facing the park with, at that time,
two million visitors each year. They describe, from forty
years combined experience working at Grand Canyon, the
beauty of the natural sights and sounds and the need for quiet
and clean air. They point out the challenges of making the
park available to the growing numbers of visitors while still
fulfilling the duty of the National Park Service to protect all
things in the park. The following short excerpt from the book
describes why we need to meet those challenges which are
even greater today. "That which enveloped us consisted of
three fundamentals: the Canyon, the sky – a deep blue, pure
empty, all-enclosing sky – and the stillness." And how many
visitors does the Canyon have today? And how many cars?
And how many scenic flights?*

107. TRIMBLE

BLESSED BY LIGHT. VISIONS OF THE COLORADO PLATEAU.
Edited by Stephen Trimble with a foreword by Edward
Abbey and an introduction by Gibbs M. Smith. Layton:
Gibbs M. Smith. 1986. Pp. xvii, 69; ills.; afterword,
photographing the plateau; contributing writers;
contributing photographers.

¶ *This is a book of photographs of the Colorado Plateau
taken by some of the best photographers that ever set up a
camera and tripod on slickrock. Along with these photographs
are excerpts from some of the best writers who give us "word
pictures" of the Colorado Plateau. Several of those writers*

are in Farquhar and several are in this bibliography. Trimble
has put together a grand combination.

108. VERKAMP

History of Grand Canyon National Park. By
Margaret M. Verkamp with a preface to the reprint by
Ronald W. Werhan and a biographical sketch by Paul
Sweitzer. Flagstaff: Grand Canyon Pioneers Society
Collectors Series Volume 1. 1993. Pp. 57; bibliography;
index; paperback.

¶ *This is a concise, accurate and well written history of
Grand Canyon National Park. It was written as a Master
of Arts thesis for the University of Arizona in 1940 and
brought out of obscurity for this edition by Mr. Werhan and
the Grand Canyon Pioneers Society [now Grand Canyon
Historical Society]. The bibliography will give the reader a
feel for the thoroughness of Verkamp's research.*

109. WAMPLER

a

Havasu Canyon. Gem of the Grand Canyon. By Joseph
Wampler. Berkeley: Joseph Wampler with chapters by Dr.
Harold C. Bryant and Weldon F. Heald and photographs
and a foreword by the author. 1959. Pp. 121; ills.; end paper
maps; bibliography; paperback.

b

Havasu, A Canyon Home. By Joseph Wampler. Berkeley:
Joseph Wampler with chapters by Dr. Harold C. Bryant
and Weldon F. Heald and photographs and a foreword
by the author. 1981. Pp. 121; ills.; end paper maps;
bibliography; paperback.

¶ *For years the author operated* Wampler Trail Trips, *a service that took tourists down to Supai Village and to the canyon and waterfalls. He knew the people and the area very well and has shared much of his knowledge in these publications. The text in both books is the same, as are almost all of the photographs. Each has a few color plates not found in the other. The first edition,* a, *is printed on a better quality paper and as a result the black-and-white photographs have a little better contrast. In* b *there are "Some Notes From Supai Tourist Information" not found in* a. *These notes are very interesting though much of this information is no longer current.*

110. WATKINS

a

STONE TIME. SOUTHERN UTAH: A PORTRAIT AND A MEDITATION. By T. H. Watkins with photographs by the author and with a preface by Terry Tempest Williams. Santa Fe: Clear Light Publishers. 1994. Pp. 104: ills.

b

THE REDROCK CHRONICLES. SAVING WILD UTAH. By T. H. Watkins with photographs by the author. Baltimore: The Johns Hopkins University Press. 2000. Pp. 163; ills.; maps; suggested readings.

¶ *Although these are totally different books in format and text they have enough in common to be listed together. Pointing out the scenic and environmental treasures of Southern Utah is the theme of both and several of the same photographs are used in each. When Watkins wrote* a, *he was editor of "Wilderness," the magazine of The Wilderness Society. In* a *he deals with the reasons many people think much of this area should be set aside as protected wilderness. The idea of the wilderness ethic, reminiscent (and credit is given) of Aldo Leopold, is introduced and carried throughout the text. In* b *Watkins deals with the efforts of the Southern Utah Wilderness Alliance and other national and regional organizations to have a large portion of Southern Utah set aside, and the reasons for those efforts. In* b *he also presents the viewpoints of the people whose families have lived in Southern Utah for several generations. Watkins fell in love with the canyon country and this theme is common to both books. "Red Rock Chronicles" was Watkins's last book.*

111. WATKINS

THE GRAND COLORADO. THE STORY OF A RIVER AND ITS CANYONS. By T. H. Watkins and contributors with a foreword by Wallace Stegner and color photographs by Philip Hyde. The contributors are William E. Brown, Jr., Robert C. Euler, Helen Hosmer, Roderick Nash, Roger Olmsted, Wallace Stegner, Paul S. Taylor, and Robert A. Weinstein. Palo Alto: American West Publishing Company. 1969. Pp. 310; ills.; end paper maps; maps; appendix; acknowledgments; notes on the authors and contributors; picture credits; index.

❡ *The text of this comprehensive story of the Colorado River is divided into three parts: The Myth, The Conquest, and The Legacy. Each details specific activities that made a significant contribution to the history of the Colorado River and the canyon country. The many historic photographs, some never before published, round out and add clarity to the text. This one volume probably has more river and canyon history than any other published work.*

112. ZWINGER

DOWNCANYON. A NATURALIST EXPLORERS THE COLORADO RIVER THROUGH THE GRAND CANYON. By Ann Haymond Zwinger with drawings by the author. Tucson: The University of Arizona Press. 1995. Pp. viii, 318; ills.; maps; appendix; notes; acknowledgments; index.

¶ *During her many trips in each of the four seasons, Zwinger explored the main canyon and many of the side canyons. She describes her experiences, many while she was alone, with an understanding of the natural world that we find in all of her writings about the canyon country. Anyone making plans to go down the river through Grand Canyon will see and understand the River and the Canyon better if they have read this book before they go.*

113. ZWINGER

RUN RIVER RUN. A NATURALIST'S JOURNEY DOWN ONE OF THE GREAT RIVERS OF THE WEST. By Ann Zwinger with illustrations and maps by the author. New York: Harper & Row, Publishers. 1975. Pp. viii, 317; ills.; maps; notes; index.

¶ *Geology, natural history, human history, and a sincere appreciation and understanding of the river and the country through which it runs, are included in this informative look at the Green River. Zwinger has covered the entire river either on foot, by canoe, raft, or from the air. This is a wonderful opportunity for an armchair exploration of the longest branch of the Colorado River from its source to the Confluence. After reading about it you just might want to go do the real thing.*

114. ZWINGER

WIND IN THE ROCK. By Ann Zwinger. New York: Harper & Row, Publishers. 1978. Pp. 258; ills.; maps; notes; acknowledgments; index.

¶ *In southeastern Utah there are five canyons draining into the San Juan River from the Grand Gulch Plateau. The largest and longest of these is Grand Gulch. The others are Johns Canyon, Slickhorn Canyon, Steer Gulch, and Whirlwind Draw. Zwinger has explored all of them, mostly on foot, and brings us accurate and beautiful descriptions of these canyons. She explores and describes both the natural history and the human history of the canyons. It will make you want to go and see for yourself.*

PART VI

GEOLOGIC STUDIES

115. BAARS

a

RED ROCK COUNTRY. THE GEOLOGIC HISTORY OF THE COLORADO PLATEAU. By Donald L. Baars. Garden City: Doubleday/Natural History Press. 1972. Pp. 264; ills.; maps; bibliography; index.

b

THE COLORADO PLATEAU. A GEOLOGIC HISTORY. By Donald L. Baars. Albuquerque: University of New Mexico Press. 1983. Pp. 279. ills.; maps; bibliography; index.

¶ *This is the basic geologic history of the Colorado Plateau written by a professional geologist who later became a research professor of geology. The clearly written text is understandable to the layman but technical enough to be appreciated by the trained geologist. One chapter includes a raft trip from Lee's Ferry to Bright Angel Creek. The titles are different but with the exception of a few minor changes in text and illustrations the two volumes are the same.*

116. BAARS/BUCHANAN

THE CANYON REVISITED. A REPHOTOGRAPHY OF THE GRAND CANYON 1923-1991. By Donald L. Baars and Rex C. Buchanan with rephotography by John R. Charlton. Salt

Lake City: University of Utah Press. 1994. Pp. 167. ills.; maps showing photograph locations; bibliography; pictorial paperback.

¶ *In 1923 Claude Birdseye led a three-month expedition down the Colorado River from Lee's Ferry through Marble and Grand Canyon. This was the last official government survey before any dams with their reservoirs were built on the Colorado River system in or above Grand Canyon. The purpose was two-fold. First, even though Powell and Stanton had both done survey work in the canyons there were still no accurate maps of the Canyon floor. Second, Birdseye was locating potential dam sites. The official photographer and hydraulic engineer was E. C. LaRue. Emery Kolb was head boatman and did some of the photography. Lewis Freeman was a boatman and also did some photography. Freeman's book, 'Down The Grand Canyon', [Farquhar #54] tells the story of the trip and his article in the May 1924 "National Geographic Magazine" gives a shorter version. The 1991 group, headed by Baars and Buchanan and photographer Charlton, was with the Kansas Geological Survey. They were able to locate photographic stations and rephotograph forty-five of the hundreds of views from the 1923 expedition. It is interesting to see by comparing photographs how much or how little change has taken place.*

117. BARNES

UTAH CANYON COUNTRY. By F. A. Barnes with a foreword by Ted Wilson. Salt Lake City: Utah Geographic Series, Inc. 1986. Pp. 117, ills.; maps; paperback.

¶ *The part of the Colorado Plateau contained in southeastern Utah is big enough and rugged enough that it would take several lifetimes to see it all. Fran Barnes has lived in Moab for half a lifetime and has, with his wife Terby, probably seen more of southeastern Utah than almost anyone. In this book*

88

he has written with a great deal of expertise about the geology and the natural and human history, with special emphasis on the national parks, national monuments, recreation areas, and wilderness areas contained in that part of the plateau. The work of many well-known photographers, including the author, is used with the text. It is a book to be read before you go, while you are there, and after you are home. There was a revised printing issued in hardback in 1987 and unless there were minor changes in the text, the only difference is the picture of Kolob Arch in Zion National Park.

118. BEUS/MORALES

a

GRAND CANYON GEOLOGY. Edited by Stanley S. Beus and Michael Morales with a foreword by Edwin H. Colbert. New York: Oxford University Press, Inc. and Museum of Northern Arizona Press. 1990. Dedicated to the memory of Dr. Edwin Dinwoodie McKee. Pp. x, 518; ills.; maps; bibliography; index.

b

GRAND CANYON GEOLOGY. Edited by Stanley S. Beus and Michael Morales with a foreword by Edwin H. Colbert. New York: Oxford University Press, Inc. and Museum of Northern Arizona. 2002. second edition. Dedicated to the memory of Dr. Edwin Dinwoodie McKee. Pp. x, 432; ills., maps; bibliography; index.

¶ *More than twenty different geologists, all proficient in their special field, have collaborated on this in-depth study of the geology of Grand Canyon. Many earlier ideas about the geologic history of Grand Canyon have evolved over the years. As the authors of this volume indicate, some of the present thinking may change in the future as new ideas are brought forward with new discoveries. This book represents*

*the latest studies in each of the specialized fields represented
in the text. This is a book for the reader that has some
background in geology in general, and of Grand Canyon in
particular. Time is the most difficult part to understand. Two
new chapters have been added in b and all chapters have
been updated where necessary. Also in b all photographs have
been replaced or re-screened for better resolution.*

119. DUFFIELD

VOLCANOES OF NORTHERN ARIZONA. SLEEPING GIANTS
OF THE GRAND CANYON REGION. By Wendell A. Duffield.
Photographs by Michael Collier. Grand Canyon: Grand
Canyon Association. 1997. Pp. 68; ills.; maps; glossary;
index; paperback.

¶ *This little book is technical enough for the volcanologist
but understandable to the layman. The aerial photographs
make many of the features very obvious that can't be seen well
from the ground. Road logs are included to help anyone with
the time and interests find most of the main features of this
unusual and amazing landscape.*

120. FOUR CORNERS GEOLOGICAL SOCIETY

CANYONLANDS COUNTRY. A GUIDEBOOK OF THE FOUR
CORNERS GEOLOGICAL SOCIETY 1975. By Four Corners
Geological Society Eighth Field Conference – September
22-25, 1975. James E. Fassett, editor with a foreword by
Edward Abbey. np: Four Corners Geological Society. 1975.
Pp. v, 281; ills.; endpaper maps, maps; advertisements;
road logs; appendix.

¶ *There is much information here in one volume. Topics include history, archaeology, geomorphology, stratigraphy and paleontology, structural geology, and economic geology. There are sub-sections under each topic written by experts and references are listed. There are three road logs. The first day is from Moab to Dead Horse Point, Grand View Point, Upheaval Dome, and the Shafer Trail. The second day is from Moab to Natural Bridges via the Needles, Dugout Ranch, and Elk Ridge and the third day is from Natural Bridges to Elaterite Basin via White Canyon and Powell reservoir. Good information. These people went over the routes.*

121. FOUR CORNERS GEOLOGICAL SOCIETY

GEOLOGY AND NATURAL HISTORY OF THE GRAND CANYON REGION. FIFTH FIELD CONFERENCE POWELL CENTENNIAL RIVER EXPEDITION 1969. By Four Corners Geological Society. np: Four Corners Geological Society. 1969. Pp. 212; ills.; maps; advertisements; river log, geologic map, and fence diagram of the Kaibab Formation in pocket.

¶ *As with the book preceding this one there is much information here in one volume. Topics include prehistory, human history, national park history, natural history, as well as geology. Each section is written by a respected expert in the field and has references cited or a bibliography for that section. The last section in the book is a road log from Yaki Point to Lee's Ferry via Cameron, Arizona. There is a river log in pocket.*

122. NABHAN/WILSON

CANYONS OF COLOR. UTAH'S SLICKROCK WILDLANDS. By Gary Paul Nabhan and Caroline Wilson and featuring the photography of Jeff Garton. New York:

91

HarperCollinsPublishers. 1995. Pp. 132; ills.; map and end paper maps; acknowledgments; index; selected bibliography by chapter.

¶ *Geology explained with text, charts, diagrams, time lines, and photographs. Gary Nabhan is a displaced flatlander from the Midwest who first saw the slickrock country at age seventeen. Caroline Wilson literally grew up in Arches National Monument, as she is one of the daughters of Bates Wilson, Superintendent of Arches from 1949 to 1971. How the geology helps to determine the life forms in the slickrock country is explained very nicely and many photographs supplement this effort. Some of the interesting features of this book are the remembrances written into the text. These are about experiences, some involving personal learning opportunities from which you, especially on your first or second visit to the area, might benefit.*

123. RABBITT/MCKEE/HUNT/LEOPOLD

THE COLORADO RIVER REGION AND JOHN WESLEY POWELL. By Mary C. Rabbitt, Edwin D. McKee, Charles B. Hunt and Luna B. Leopold with a foreword by W. T. Pecora. Washington: United States Government Printing Office. 1969. Geological Survey Professional Paper 669. Pp. xi, 145; ills.; maps; notes; references; bibliography.

¶ *Each of the four authors has written a paper in honor of Powell on the hundredth anniversary of his exploration of the Colorado River. Mary Rabbitt has contributed "John Wesley Powell: Pioneer Statesman of Federal Science." She gives a good history of science in America as Powell would have known it, followed by some of his contributions. Edwin McKee has written on "Stratified Rocks of the Grand Canyon." He tells of Powell's pioneering contributions to geology in this area and something of the knowledge gained since Powell's time. Charles Hunt's contribution is titled "Geologic History of the Colorado River." He points out that Powell had many things correct in his hypothesis and some of the people that worked with him built on and refined his theories. Luna Leopold wrote on "The Rapids and Pools – Grand Canyon." He explains why rapids are where they are and why there are different forms of rapids. Altogether this is a notable addition to the writings of Grand Canyon and the Colorado River.*

124. REDFERN

CORRIDORS OF TIME. 1,700,000,000 YEARS OF EARTH AT GRAND CANYON. Panoramic photography and text by Ron Redfern with illustrations by Gary Hincks and an introduction by Carl Sagan. New York: The New York Times Company. 1980. Pp. 198; ills.; maps; acknowledgments; bibliography; index.

¶ *This is a book about geology, but not "just another book about geology." This is a book with great photographs of Grand Canyon and other scenes on the plateau, but not "just another book with great photographs." This is a very good geology book with unique panoramic photographs, some with up to 160 vertical degrees and 320 horizontal degrees of view. The human eye cannot see that much without moving the head. This is one of the first books for the non-geologist that goes into a good explanation of continental drift and plate tectonics and Redfern includes accurately rendered illustrations to show these processes. Many of the panoramic photographs are taken from places very seldom visited and difficult to reach. You will have greater knowledge and appreciate the natural processes at work in Grand Canyon and on the Colorado Plateau after spending time with this book. In 1980 there was an identical printing published by Orbis of London and in 1983 Reader's Digest printed an edition of this title with the same contents in a 10" square format rather than the original 12" square.*

93

125. STEPHENS/SHOEMAKER

IN THE FOOTSTEPS OF JOHN WESLEY POWELL. AN ALBUM OF COMPARATIVE PHOTOGRAPHS OF THE GREEN AND COLORADO RIVERS, 1871-72 AND 1968. By Hal G. Stephens and Eugene M. Shoemaker. Foreword by Bruce Babbitt. Boulder: Johnson Books and Denver: The Powell Society, Ltd. 1987. Pp. x, 285; ills.; maps; tables; glossary; selected reading; camera station index; index.

¶ *Major Powell's first expedition down the Green and Colorado Rivers did not include a photographer. Three different photographers served the second expedition: E. O. Beaman, James Fennemore, and Jack Hillers (see #45 for this tale). Most of the photographs from the second Powell expedition have never before been printed in album or book format. This publication remedies that situation. Some of the photographs were used to make line drawings which are found in early publications, Powell's included. Many of them were used as stereophotographs but very few of those are available today. Due to the flooding of Glen Canyon by the reservoir many camera stations are no longer available. The authors did identify about 150 camera stations from Green River, Wyoming down to Kanab Creek and 110 of those were selected for this publication. As in all rephotography projects, many of the photographs show little if any change. The real strength of this book is that the photographs have been made available again.*

126. WEBB

GRAND CANYON, A CENTURY OF CHANGE. REPHOTOGRAPHY OF THE 1889-1890 STANTON EXPEDITION. By Robert H. Webb. Tucson: The University of Arizona Press. 1996. Pp. xv, 290; ills.; maps; figures; appendix; notes; index.

¶ *For his river-level railroad survey Robert Brewster Stanton wanted photographs made approximately every mile. Franklin A. Nims was his photographer until he fell from a cliff in upper Marble Canyon and had to be taken out, which was no small task. Stanton then assumed the photographic responsibilities, never having used a camera before. From 1989 to 1995 Robert Webb matched Stanton's camera stations and time of day as closely as possible and rephotographed all 445 of the Stanton Grand Canyon scenes. Forty-five of those scenes have been reproduced in this book. By comparing the old and new photographs, the impact Glen Canyon Dam has had on sandbar erosion and the increase in riparian vegetation (much of it tamarisk) becomes obvious. The author is a USGS hydrologist and in the text gives a comparative analysis of the matching photographs. Some of the rapids have changed little if any in the hundred years between photographs while some are hardly recognizable as the same rapid. Some examples of vegetation are quite remarkable. In one of the Stanton photographs a dead juniper is shown and in the later view the same dead juniper is still there, almost unchanged. Chapter 1 gives a good account of Stanton's work in the canyons. Other chapters discuss plant and animal life in the Canyon and the effects of floods. The last chapter discusses how the canyon experience changed Stanton. After he became photographer he became interested in the scenery, not just the river-level route for the railroad. People can change the Canyon but the Canyon can also change people.*

Part VII

THE LOWER RIVER AND THE SALTON SEA

127. BRANYON/CHAWKINS/FRIZIER/ JOHNSON

COLORADO RIVER RECREATION GUIDE. AN ATLAS OF THE COLORADO RIVER FROM GRAND LAKE, COLORADO TO YUMA, ARIZONA. By Max Branyon, Steve Chawkins, Deborah Frizier, Carolyn R. Johnson, and others. Denver: Aquamaps, Inc. 1986. Pp. 1-27, 2-16, 3-31, 4-16, 5-19, 6-22, 7-31, 8-19, 9-26, 10-25. ills; maps; paperback.

¶ *This folio volume covers a larger territory but is reminiscent of Erickson [Farquhar #95]. Reading from its cover, the atlas includes: "Colorado River History, Upper Colorado River Whitewater, Lake Powell, Lake Mead, Lake Mojave, Lake Havasu, Parker Strip, Eleven Principal Reservoirs of the Lower Colorado River and the Major Arizona Reservoirs. An accurate, informative, easy to use guide for water sports enthusiasts." The maps are excellent and the text is very informative with many historic photographs. Much of the service directory and campground information may no longer be accurate but some is still useful.*

128. COOK

a

LEGENDS OF THE LOWER COLORADO. By Fred S. Cook. Volcano: The California Traveler. 1973. Pp. 72. ills.; advertisements; paperback.

b

HISTORICAL LEGENDS OF THE LOWER COLORADO. Fred S. Cook, editor. Parker: The River Reporter. 1989. Pp. 70. ills; advertisements; paperback.

¶ *An entertaining and informative collection of stories about early explorations, steamboats, mining towns, mines and miners, developers, bad men, the Parker Dam War, the Salton Sea, and many other topics. One of the most interesting features is the collection of historical photographs, several of which probably have not been published before. The format in both titles is much alike but very little information is duplicated. The cover on b indicates it is Vol. I but as Mr. Cook passed away in 1989 additional volumes were probably never published.*

129. CROWE/BRINCKERHOFF

EARLY YUMA. A GRAPHIC HISTORY OF LIFE ON THE AMERICAN NILE. Edited by Rosalie Crowe and Sidney B. Brinckerhoff. Flagstaff: Northland Press. 1976. Pp. vii, 135. ills.; maps.

¶ *After an excellent introduction by Rosalie Crowe, this volume covers the history of Yuma, Arizona from the 1850s into the 1880s using, for the most part, excerpts from newspapers of the time. Photographs illustrating the news articles or similar events are placed with the articles. After finishing the book you should have a good feel for what it might have been like to visit or live in Yuma during any of that period. Remember, these people didn't have running water or air conditioning!*

130. DeBUYS

SALT DREAMS. LAND & WATER IN LOW-DOWN CALIFORNIA. By William deBuys with photographs by Joan Myers. Albuquerque: University of New Mexico Press. 1999. Pp. xiii, 307. ills.; maps; notes; references; index.

¶ It is natural for water to run downhill. Starting again in late winter of 1905, due mostly this time to man's fiddling with the environment, the Colorado River ran downhill to the Salton Sink. It ran for two years. This large, handsome volume is a modern history of what we know today as the Salton Sea, the expectations that resulted from the fact that it was there, and the disappointments when those expectations were not realized. It is a fascinating and in many ways a sad story. Joan Myers's black-and-white photographs catch the mood of the story exceptionally well. This book is a noteworthy addition to the history of the lower river.

131. DEKENS

RIVERMAN DESERTMAN. By Camiel Dekens as told to Tom Patterson. Riverside: Press-Enterprise Company. 1962. Pp. 111; ills.; paperback.

¶ Camiel Dekens came to the Palo Verde Valley in the fall of 1907 at the age of 20. At this time the town of Blythe, California had not yet been founded and Ehrenberg, Arizona, across the Colorado River, still had two saloons. In the early 1900s the valley was used mostly by cattlemen but the first homesteaders were starting to come in and settle. At this time the idea of irrigation for the farms was becoming popular. There are stories of mines, experiences on the river, cattle rustling, road building, fist fights, water companies, and lots of hard work. Dekens had a part in many of these activities and related them to Tom Patterson, who has set them down to make a very readable and informative little book.

132. FORBES

WARRIORS OF THE COLORADO. THE YUMAS OF THE QUECHAN NATION AND THEIR NEIGHBORS. By Jack D. Forbes. Norman:

The University of Oklahoma Press. 1965. Pp. xx, 378. ills.; maps; appendix; glossary; bibliography; index.

¶ *The first contact the Native Americans living along the lower Colorado River had with Europeans was most likely in 1540 when Alarcón and Díaz were in the area. They were along the river because of their association with the expedition lead by the man we know as Coronado. This study, one of the volumes from the excellent American Indian Series of the University of Oklahoma, gives a complete account of these tribes and the important part they played in the history of the Lower Colorado River and the far Southwest.*

133. IVES

STEAMBOAT UP THE COLORADO. FROM THE JOURNAL OF LIEUTENANT JOSEPH CHRISTMAS IVES, UNITED STATES TOPOGRAPHICAL ENGINEERS, 1857-1858. Edited by Alexander L. Crosby, illustrated by Lorence Bjorklund with a foreword by the editor. Boston: Little, Brown and Company. 1965. Pp. 112; ills.; map; index.

¶ *Yes, Ives is Farquhar #21. This book is primarily for the younger reader but those of all ages can enjoy it, especially since an original Ives is difficult to find, very expensive, and reprints are not common. Crosby has abridged the Ives journal but the essence of the Lieutenant's experience is still here.*

134. LEOPOLD

a

ROUND RIVER. FROM THE JOURNALS OF ALDO LEOPOLD. By Aldo Leopold. Edited and with a preface by Luna B. Leopold and illustrations by Charles W. Schwartz. New York: Oxford University Press. 1953. Pp. 173; ills.

99

b

ROUND RIVER. FROM THE JOURNALS OF ALDO LEOPOLD.
By Aldo Leopold. Edited and with prefaces to the 1953
edition and the 1991 edition by Luna B. Leopold and
illustrations by Mary A. Schafer. Minocqua: NorthWord
Press, Inc. 1991. Pp. 248; ills.

¶ *Only one chapter of this book deals with the Colorado River
but it tells a significant story. In the fall of 1922 Aldo Leopold
and his brother Carl spent almost three weeks in the Colorado
Delta. This is a good description of the delta in its natural
state when the river still ran to the sea. They were there to hunt
and at that time there were many things to hunt. They shot at
most anything that moved. They were young and game was
plentiful. This was before Leopold helped shoot an old she-wolf
below a high rim rock in the southwest and got to her "in time
to watch a fierce green fire dying in her eyes" that he writes
about so eloquently in his classic, "A Sand County Almanac."
Both printings contain the same text. The type was reset for* b
making it easier to read.

135. MARTIN

YUMA CROSSING. By Douglas D. Martin with illustrations
by Horace T. Pierce. Albuquerque: The University of New
Mexico Press. 1954. Pp. ix, 243; ills.; map; bibliography;
index.

¶ *This is not a history of Yuma, Arizona but a history of the
place where people crossed the Colorado River. Rather late
in the history of that crossing, the town of Yuma was built.
Europeans had used the crossing for at least four hundred
years when this book was written and Native Americans had
crossed there for hundreds of years before that. The Spanish
conquistadors, the mountain men, the Army of the West, the*

Mormon Battalion, the gold hunters, settlers, the railroad builders, and countless others were funneled to this crossing. Martin writes about all of them, and tells their stories in chronological order.

136. ODENS

FIRE OVER YUMA. TALES FROM THE LOWER COLORADO. By Peter Odens with photographs by the author. Yuma: Southwest Printers. 1966. Pp. 59; ills.; paperback.

¶ *This book contains interesting and well written stories of early happenings along the Colorado near Yuma, Arizona. In stories of the military Odens relates Whipple's second visit, when the Quechan chief's daughter recognized the Lieutenant and averted possible trouble. He also reports Ives's trip with the steamboat* Explorer *and the later discovery of the boat's remains in the sand near the delta. One of the more interesting yarns includes John Cremony's comment to the Indians that his telescope could "kill the moon." John Glanton's short career at the ferry is mentioned as is the territorial prison at Yuma with a tale about Pearl Hart, perhaps its most famous inmate.*

137. PAHER

CALLVILLE. HEAD OF NAVIGATION ARIZONA TERRITORY. Edited by Stanley W. Paher with etchings and illustrations by Roy E. Purcell. Las Vegas: Nevada Publications. 1981. Pp. 24 plus "On The Banks of the Mother River" by Roy E. Purcell; ills.; maps; pictorial paperback.

¶ *Located just above Black Canyon in the bend where the Colorado is turning from a westerly to a southern direction, Callville was the turn-around point for the early steamboats hauling freight up river. It was often a muddy reddish colored river then. Now the approximate location is a harbor for power and sail boats on the clear reservoir above Hoover Dam. This is a good collection of the history of Callville from its founding in December of 1864 until its disappearance in 1935 under the waters of the reservoir.*

138. PAHER

a

COLORADO RIVER GHOST TOWNS. By Stanley W. Paher in collaboration with Robert L. Spude with etchings and illustrations by Roy E. Purcell. Las Vegas: Nevada Publications. 1976. Pp. 63 plus "On The Banks of the Mother River" by Roy E. Purcell; ills.; map; hints for desert travelers; for further reading.

b

COLORADO RIVER GHOST TOWNS. By Stanley W. Paher in collaboration with Robert L. Spude. Las Vegas: Nevada Publications. 1976. Pp. 48; ills.; map; hints for desert travel; for further reading; paperback.

❡ *Some were river ports, some were mining towns, and at least one was a "blow-off" town for the troops from Fort Mohave. Many of the individual histories of these towns are only one paragraph long but that is about all the history they had. A vein of silver or gold or copper could start a town but when the vein petered out, and most did very soon, everyone left for the next "boom town" to make their fortune. A very few such as Chloride, Oatman, and Searchlight have survived in some form, but the glitter of the early days is all gone. The old photographs collected here help to make this book very desirable. Included in* a *are some towns and photographs not found in* b. *Also not included in* b *is the art work and poetry that are found in the section "On the Banks of the Mother River." Two hundred copies of* a *were signed and in slipcase.*

139. ROBINSON

"Wife at Port Isabel. A Pioneer Woman's Colorado River Letters." By Ellen Robinson, edited by Frank S. Dooley, M.D. The Westerners Brand Book Los Angeles Corral Book #7. Los Angeles: The Los Angeles Westerners. 1957. Pp. 271-285; ills.

¶ *Imagine if you can that you are a young lady living in Maryland in 1869. You have never been more than fifty miles from home. After a brief courtship you marry a man many years your senior and almost immediately after your wedding leave on a train for the west coast. You eventually end your journey by the mouth of the Colorado River at the shipyard and settlement of Port Isabel. At this time it would have been difficult to choose a more inhospitable place to live on the continent of North America. These are the letters from Ellen Robinson, the new bride, to her family at home. In these letters she describes, sometimes in great detail, the surroundings and conditions where she is living. She writes of being left alone by her husband, a Colorado River steamer captain and pilot, of their various illnesses, and about her homesickness. She writes home of the joy of having a child and of earthquakes along the Colorado River. Robinson's Maryland ancestors would certainly have been proud of her as she carried the same pioneering genes.*

140. WOODWARD

Feud on the Colorado. By Arthur Woodward. Los Angeles: Westernlore Press. 1955. Pp. xiii, 165; ills.; bibliographical notes; index.

¶ *The feud was between George Alonzo Johnson, Captain of the steamboat General Jesup, and Lieutenant Joseph Christmas Ives and his steamboat Explorer. In a race of any kind, both competitors can't be first, but in the race to see who would be first to reach the head of navigation up the Colorado River above Yuma, both of these gentlemen claimed the honor. Woodward researched all the records about this event and rediscovered the report of Lieutenant J. L. White who was aboard the General Jesup on its northbound trip. There is a lot of river history in this very interesting story. This is the last title in Donald M. Powell's little bibliography, "Arizona Fifty," and his comment about Woodward's book is still appropriate: "Who won? Read the book and find out."*

Part VIII

GLEN CANYON

141. CRAMPTON

a

GHOSTS OF GLEN CANYON. HISTORY BENEATH LAKE POWELL. By C. Gregory Crampton with a foreword by Lyman Hafen and a prefatory note by C. Gregory Crampton. St. George: Publishers Place, Inc. 1986. Pp. 135; ills.; maps; bibliography; index of proper names; paperback.

b

GHOSTS OF GLEN CANYON. HISTORY BENEATH LAKE POWELL. By C. Gregory Crampton with an introduction by Will Rusho and a prefatory note by C. Gregory Crampton. Salt Lake City: Cricket Productions. 1994, revised. Pp. 135, plus C1-C16 containing color photographs by Will Rusho; ills.; maps; bibliography; index of proper names; paperback.

¶ *Man has lived and worked in Glen Canyon for thousands of years but a peak of activity erupted in the late 1800s with a gold rush. Crampton spent a good part of six years performing historical survey work prior to the flooding of Glen Canyon by the reservoir. This volume is one of the results of that effort. There are brief discussions of the various historical periods prior to the gold rush but most of the book is devoted to the ruins and relics of the period from 1883 to the end of the "boom" in 1912. The back cover of a has a brief note about Dr. Crampton and indicates he is retired and living in St. George, Utah. There is also a five line quote on the back cover from Edward Abbey.*

Printing b *has the same note about Dr. Crampton and a three line quote on the back cover from Stan Jones. There was also a 1994 printing by Tower Productions; in the introduction and on the back cover mention is made of Dr. Crampton's death. Internally all three printings are identical except for the 16 pages of color plates that are not in* a. *If you would care to read more of Dr. Crampton's writings on the history of Glen Canyon look in University of Utah Anthropological Papers Numbers 42, 46, 54, 61, 70, and 72.*

142. GASKILL

PEACEFUL CANYON GOLDEN RIVER. A PHOTOGRAPHIC JOURNEY THROUGH FABLED GLEN CANYON. Compiled by David and Gudy Gaskill and with a foreword by Bruce Berger. Golden: The Colorado Mountain Club Press. 2002. Pp. 95 plus CD-ROM; ills.; maps; photographers' credits; writers' credits; paperback.

¶ *The Gaskills have collected pictures from many sources, taken on pre-dam float trips. They have put them in the order you would see the scenes if you were coming down the river from Hite, about where backwater on Powell Reservoir is now, to Lee's Ferry. All the photographs are in color. With the pictures are quotes from journals, diaries, and books that describe the scenes or similar scenes. Included with the book is a CD-ROM that contains over eight hundred additional photographs of Glen Canyon with the same kind of information as is in the book proper. Does this CD-ROM addition give us a preview of books of the future?*

143. HYDE

A GLEN CANYON PORTFOLIO. By Philip Hyde. Flagstaff: Northland Press, 1979. 20 unmounted black-and-white photographs, image size, 9-1/2″ x 12″. A 2 page

"Photographer's Comment" is with the photographs. In the same folder is THERE WAS A RIVER, by Bruce Berger. Flagstaff: Northland Press, 1979. Pp. 12 folio; paperback.

¶ *These black-and-white images were taken from 1955 through 1964 and with Hyde's expertise with a camera nothing needs to be said about the quality of the work. Hyde's eulogy to Glen Canyon is given in his "Photographer's Comment" and his skill with the pen is almost equal to his skill with the camera. Bruce Berger's story is about a trip taken down Glen Canyon in 1962 with Katie Lee as trip leader. Who would want a more qualified guide? It was first published in "Mountain Gazette," No. 31, 1975. This was one of the last trips taken down the river through Glen Canyon before the gates of the dam were closed.*

In 1994 the University of Arizona Press published a book of Bruce Berger's essays with the same title as his article in this portfolio. The first sixty pages are a story about the same river trip but it has been almost totally rewritten from a perspective of thirty years later. It is interesting to compare the two versions.

144. INSKIP

THE COLORADO RIVER THROUGH GLEN CANYON BEFORE LAKE POWELL. HISTORIC PHOTO JOURNAL 1872-1964. Edited by Eleanor Inskip. Moab: Inskip Ink and Page: Glen Canyon Natural History Association. 1995. Pp. ii, 95; ills.; map; photography credits; journal credits; paperback.

¶ *No doubt Inskip's sixteen years with the Canyonlands Natural History Association contributed to her goal of putting this volume together. From all the sources available to her she selected 101 images by thirty different photographers, both professional and very accomplished amateurs, and journal entries from forty different writers. Most, but not all, of the journal entries have been published before. The majority of the photographs are in color and many of them have not been previously published. Each photo caption includes the Powell reservoir buoy number making it possible to find the site today. This volume is a valuable record of what Glen Canyon was and of what has been lost. There was a limited number of this printing bound in silk over boards and hand tied.*

145. LADD

LAKE POWELL. A PHOTOGRAPHIC ESSAY OF GLEN CANYON NATIONAL RECREATIONAL AREA. Photography by Gary Ladd, interpretive text by Anne Markward. Santa Barbara: Companion Press. 1994. Pp. 95; 102 photographs.

❡ *Because Glen Canyon was so beautiful the reservoir that fills it just had to be beautiful also. The photographs in this collection represent the photographer's best images from about three hundred days of backpacking and day hikes over a period of fifteen years. You will see scenes in the book that you wouldn't see as a visitor, but it might make you want to park your boat and use your feet in a few places.*

146. LEE

ALL MY RIVERS ARE GONE. A JOURNEY OF DISCOVERY THROUGH GLEN CANYON. By Katie Lee, with an introduction by Terry Tempest Williams. Boulder: Johnson Books, 1998. Pp. xii, 260; ills.; maps; river miles; prologue; epilogue; annotated bibliography; acknowledgments.

❡ *This is a book about a river and a canyon but it is much more than that. Katie Lee, most often in the company of Tad Nichols and Frank Wright, both recently deceased, made many*

*leisurely trips through Glen Canyon before the dam and the
reservoir. She writes of the beauty, the quiet and the peace that
was Glen Canyon. This is one of the best verbal descriptions
of the whole mood and feeling of Glen Canyon that has been
put into print. This book is a love story, a beautifully written
love story.*

147. NICHOLS

GLEN CANYON: IMAGES OF A LOST WORLD. By Tad Nichols,
edited and with a preface by Kathleen Bryant and an
afterword by Gary Ladd. Santa Fe: Museum of New
Mexico Press. 1999. Pp. ix, 157; ills.; map; notes about
the photography.

¶ *Contrasts in texture of stone and trees and water, of light and
shadow, reflections and the curves of water and wind carved
sandstone – these were the primary features of Glen Canyon,
but the list is meaningless when you see the ensemble. The
whole is greater than the sum of its parts. The images in this
beautiful book are from approximately thirty trips over thirteen
years and were selected from more than two thousand images
of Glen Canyon and the side canyons. Many of these trips were
in the company of Katie Lee and the late Frank Wright, both
of whom contributed a short statement for the book. The group
called themselves "We Three" and they explored most, and
named many, of the side canyons. Images of Music Temple,
Hidden Passage, Twilight Canyon, Forbidding Canyon, and
many others the reservoir drowned are here. Much of the text is
taken from Nichols's journals. As beautiful as these black-and-
white images are, the photographer might have agreed they are
only reminders of what was really there. Still, this book is one
of the best remembrances we have of the lost world that was
Glen Canyon.*

148. PORTER

a

THE PLACE NO ONE KNEW. GLEN CANYON ON THE COLORADO. By Eliot Porter, edited and with a foreword by David Brower. San Francisco: Sierra Club. 1963. Pp. 169; ills. [72 plates]; endpaper maps; the Glen Canyon Community; references.

b

THE PLACE NO ONE KNEW. GLEN CANYON ON THE COLORADO. By Eliot Porter, edited and with a foreword by David Brower. San Francisco: Sierra Club. 1966. second edition. Pp. 186; ills. [80 plates]; endpaper maps; the Glen Canyon Community; references.

c

THE PLACE NO ONE KNEW. GLEN CANYON ON THE COLORADO. By Eliot Porter, edited and with a foreword by David Brower. Salt Lake City: Gibbs M. Smith, Inc., Peregrine Smith Books. 1988. Commemorative Edition 1963-1988. Pp. 180; ills. [77 plates]; endpaper maps; the Glen Canyon Community; references.

d

THE PLACE NO ONE KNEW. GLEN CANYON ON THE COLORADO. By Eliot Porter, edited and with a preface by David Brower. Salt Lake City: Gibbs M. Smith Publisher, 2000 "Commemorative Edition, the beginning of a century," in cooperation with the Glen Canyon Institute. Pp. 192; ills. [77 plates]; afterword; the Glen Canyon Community; references; annotated bibliography of Glen Canyon; gatefold paperback.

¶ *The title of this book should probably have been "The Place Not the Right People Knew" because many people did know about Glen Canyon – several books in this list attest to that fact. This is still an important item in the history of the Colorado River in that it came at a time when the many people* and groups who opposed more dams were getting organized and speaking with a much stronger voice than in the past. They realized Glen Canyon was lost and didn't want to lose anything else. Basically a and b are the same except for the number of plates. The printings in c and d are in a slightly smaller format and d has no maps. The afterword in d, written by Daniel P. Beard, former Commissioner of the Bureau of Reclamation, just might surprise you. You will certainly find it of interest. Near the back of the book d has a three-page section written by David L. Wegner and Pamela W. Hyde, both of the Glen Canyon Institute. As always, Eliot Porter's images speak for themselves.

149. RUSSELL

ON THE LOOSE. By Terry and Renny Russell. San Francisco: The Sierra Club. 1967. Pp. 122; ills.

¶ *Relatively few pages of text and photographs are along the Colorado River, but those that are relate to Glen Canyon. The insights of these two young men are to be envied by many older heads. They include some appropriate quotes from several writers that most readers of this work will recognize. All the text is written in calligraphy, done by one of the co-authors, and most of the pictures were taken by the authors. Be sure to spend a few minutes thinking about the meaning of pages 93–95. Before the book was released Terry was lost in a rafting accident on the Green River. Some hardback copies are in slip case, some in dust jackets. Also released in paperback.*

150. SPRANG

a

GOOD-BYE RIVER. By Elizabeth Sprang. Reseda: Mojave Books. 1979. Pp. 67; ills.; map.

b

GOOD-BYE RIVER. By Elizabeth Sprang with a foreword by Mark MacAllister. Las Cruces: Kiva Press. 1992. Pp. 66; ills.; map; paperback.

¶ *This is the record of a very leisurely five-week raft trip through Glen Canyon in the fall of 1959, just before the beginning of construction for Glen Canyon Dam. Sprang describes with sensitive and beautiful prose many of the very special places that have now been under water for nearly forty years. Places like Tapestry Wall, Smith Fork, petroglyph sites, the river bars where, on many of them, signs of the gold mining days of Glen Canyon were still visible. They occasionally camped in the same place for several days and sometimes encountered rain and wind and sand in their food, which, as they say, "goes with the territory." The last chapter, titled Overview, is only five pages long and can be read in less than five minutes, but many hours of thinking can come out of it. It is even more appropriate today than when it was written. With very minor changes the same text is found in both printings and there are a few differences in the sketches and photographs. The most significant difference is in format with a being octavo, available in either deluxe cloth binding or paperback. The 1992 printing, b, is oblong and available only in paperback.*

151. TELFORD/SMART

LAKE POWELL. A DIFFERENT LIGHT. John Telford, photographs and a preface, and with a preface and text by William Smart. Salt Lake City: Gibbs Smith Publisher. 1994. Pp. 95; ills.; acknowledgments; bibliography; paperback.

¶ *From the photographer's preface: "This book is not meant as a celebration of Lake Powell, nor is it a memorial to what was lost. It is a celebration of what has survived and what is still fragile and threatened." Several of the photographs are broad landscapes or lakescapes showing the enormity of the present buttes and cliffs, even though up to five hundred or more feet of them are under water. Many views are away from the reservoir in side canyons that were inaccessible before, and many of the images are in the style of Eliot Porter, one of John Telford's mentors. The last chapter of text titled "Looking Ahead," points out some of the problems that three to four million visitors each year are causing. There are solutions to these problems, but they will require an attitude adjustment for most visitors.*

152. TOPPING

GLEN CANYON AND THE SAN JUAN COUNTRY. By Gary Topping. Moscow: University of Idaho Press. 1997. Pp. xv, 404; ills.; map; acknowledgments; chapter notes; bibliography; index.

¶ *Native Americans lived in the canyons and side canyons of the Colorado and San Juan Rivers for thousands of years as their rock art and structures still attest. For several hundred years Paiute and Navajo moved along the fringes – in some cases near the rivers – and some still live in the area. The Spanish Entradas criss-crossed this area but the people did not stay. Trappers, explorers, scientists, and prospectors came but most of them did not stay long either. Finally came Indian traders, ranchers, and a few who wanted to irrigate and farm near the San Juan. The goal was to make a living and some did, for a while at least – some still do. Now the boaters and backpackers are the visitors. Dr. Topping did his homework in the libraries and archives, and interviewed descendants of the early settlers and the present residents. He also went over the country on foot, on horseback, by jeep, and by raft. All these stories and many more are in this well researched and well written addition to the literature of the Canyon Country.*

PART IX

DAMS AND DEVELOPMENT

153. ATON

INVENTING JOHN WESLEY POWELL: THE MAJOR, HIS ADMIRERS AND CASH-REGISTER DAMS IN THE COLORADO RIVER BASIN. By James M. Aton. Cedar City: Southern Utah State College. 1988. Pp. 24; paperback.

¶ *Would Powell have approved of the reservoir named for him? Would he have approved of any of the high dams built by the Bureau of Reclamation and the Corps of Engineers on the Green, the Colorado, the Gila, the San Juan, or the Gunnison Rivers? With these thoughts in mind, one of the pioneers in environmental history takes a look at the writings of John Wesley Powell. He makes some very educated guesses as to what Powell would think about the past and present policies of the Federal Government in the areas of irrigation, flood control, and power generation in the arid west. He points out how those that approve of the present policies and those who don't have both used Powell's words to justify their side of the discussion. This little volume is a printing of Distinguished Faculty Lecture No. 9, delivered by Dr. Aton at Southern Utah State College [now University] December 1, 1988. The Powell Aton finds would probably not have approved of what has been done to the Colorado River system.*

154. FARMER

GLEN CANYON DAMMED. INVENTING LAKE POWELL & THE CANYON COUNTRY. By Jared Farmer. Tucson: The University of Arizona Press. 1999. Pp. xxvii, 269; ills., maps; notes; suggestions for further reading; illustration credits; index.

¶ *Farmer traces the history of the Glen Canyon area from the first "road," the Hole-in-the-Rock route, to the present ease of access to several points around and along the reservoir. The "discovery" and development of what was, in the 1930s, the largest roadless area in the lower 48 states, has happened for the most part since the construction of the Glen Canyon Dam. There have been many changes between the time of the wanderings and disappearance of Everett Ruess and the party-boat masses of today. These changes are well documented.*

155. FRADKIN

A RIVER NO MORE. THE COLORADO RIVER AND THE WEST. By Philip L. Fradkin. New York: Alfred A. Knopf. 1981. Pp. xviii, 360; ills.; maps; selected bibliography; index.

¶ *If you measure by the longest branch of the Colorado, the Green, the river is about 1700 miles from mountain snow banks in the Wind River Range of Wyoming to salt water in the Gulf of California. The shorter branch comes out of Colorado. The main river and both branches and their tributaries have been damned. The waters are used for irrigation of crops and cow pastures, hydroelectric power production, mining, drinking, watering golf courses, and several other human "needs." It is a rare year when the Colorado River is able to run far enough south to reach salt water. This book covers all parts of that story very well.*

156. FREEMUTH

ISLANDS UNDER SIEGE. NATIONAL PARKS AND THE POLITICS OF EXTERNAL THREATS. By John C. Freemuth. Lawrence: University Press of Kansas. 1991. Pp. xiv, 186; ills.; maps; appendixes; notes; interviews; index.

¶ *Although every National Park and National Monument in the system is subject to external threats, this study concentrates on "The Golden Circle of Parklands" – also known as the Four Corners Country. These threats may come in several forms such as air pollution which can lead to reduced visibility. The Navajo Generating Station is discussed at length in this regard. However, one of the main thrusts of this study is concerned with the activities that might be (or are) taking place just outside the boundaries of the parks and monuments. If you are not familiar with what are usually referred to as tar sands, or have concerns about the possible significance of tar sands, read this book. It might be particularly important to know about the large deposit just outside Canyonlands National Park and partially inside the boundaries of Glen Canyon National Recreation Area. If you want to know how boundaries are sometimes established for parks and recreational areas you probably should read this book. This is a timely study and the problems it points out will need to be re-evaluated often.*

157. GOODMAN/MCCOOL

CONTESTED LANDSCAPES. THE POLITICS OF WILDERNESS IN UTAH AND THE WEST. Edited by Doug Goodman and Daniel McCool. Salt Lake City: The University of Utah Press. 1999. Pp. xvii, 266; maps; tables; chapter notes; index; paperback.

¶ This book was the result of a University of Utah class titled "The Politics of Wilderness in Utah and the West," open to both graduate and undergraduate students. One of the class expectations to author or co-author a book chapter based on original research. The class members came from diverse backgrounds and brought a wide range of opinions. Speakers were brought in from both the pro-wilderness camp and the multiple-use sector. They took an extended field trip to San Juan County where many more people met with the class and presented different points of view. The students then went to work and the total project has probably produced as well balanced a view of the "Contested Landscapes" of the West as we are likely to see. Reading it may not change your mind but you will at least be aware of some major views and concerns of the "other side."

158. HARVEY

A SYMBOL OF WILDERNESS. ECHO PARK AND THE AMERICAN CONSERVATION MOVEMENT. By Mark W. T. Harvey. Albuquerque: University of New Mexico Press. 1994. Pp. xx, 368; ills.; maps; notes; bibliography; index.

¶ In the middle of the twentieth century there was a proposal to build two dams on the Green River within the boundary of Dinosaur National Monument. The upper dam would have backed water at least half-way up Steamboat Rock at Echo Park and higher at full flood pool. Back water would also have been well up the Green into the Canyon of Lodore and far up the Yampa. This is a complete history of the battle between the National Park Service, various conservation organizations, and preservationists on one side; and the Bureau of Reclamation, bureaucrats, and developers on the other side. It is the story of what happened after the book "This is Dinosaur," #165 in this list.

159. HUNDLEY

Dividing the Waters. A Century of Controversy Between the United States and Mexico. By Norris Hundley, Jr. Berkeley: University of California Press. 1966. Pp. ix, 266; ills.; maps; notes; bibliography; index.

¶ *Although this controversy is about three rivers, the Tijuana, the Rio Grande, and the Colorado, it is an important part of the history of use and over-use of water in the West. Most of the water for all three rivers falls on the United States, but Mexico has claims to certain amounts of the water. Unfortunately, most of the estimates of acre-feet available were made during a wet cycle. Most years all claims for the upper and lower basins and Mexico are not supplied by the watersheds. The record of the controversy for dividing the waters is all here, but unfortunately the final solution is not. Again, more people should have listened to Powell.*

160. MARTIN

A Story That Stands Like a Dam. Glen Canyon and the Struggle for the Soul of the West. By Russell Martin. New York: Henry Holt and Company. 1989. Pp. 354; maps; epilogue; acknowledgments; bibliography; index.

¶ *Martin thoroughly covers the pros and cons of the Bureau of Reclamation on one side, and the environmentalists on the other, in the building of Glen Canyon Dam. He systematically includes the background information needed to understand the philosophy of each group. Don't let the title keep you from reading this well written documentation of the controversy. This book is well researched and gives a balanced view of both sides of the issues.*

118

161. MILLER/WEATHERFORD/THORSON

THE SALTY COLORADO. By Taylor O. Miller, Gary D. Weatherford and John E. Thorson with a foreword by Max Linn and William K. Reilly. Washington: The Conservation Foundation and Napa: John Muir Institute. 1986. Pp. xv, 102. ills., map laid in; acknowledgments; executive summary; resources; index; paperback.

¶ *Stream water running through a desert will dissolve minerals from the rocks and soil. When the stream is allowed to flow naturally to the ocean the minerals, many of them salts of various kinds, add their volume to the ocean and are diluted enough to be insignificant. But that process over tremendous time is what made the oceans salty. When those streams are dammed, the water evaporates, the minerals stay, and the water behind the dam becomes saltier. When the water is run onto land to irrigate crops, the water is used by the plants or evaporates, and many of the minerals stay in the soil. After enough time this process leaves the soil too "salty" to grow crops. This report on the Salty Colorado will be an eye opener for many of you, and some won't like what this report has to say. It points out some problems and some solutions, both of which cost money. There are other sources for this information but this report has all the basic facts in an understandable, easy-to-read format.*

162. MOREHOUSE

A PLACE CALLED GRAND CANYON. CONTESTED GEOGRAPHIES. By Barbara J. Morehouse. Tucson: The University of Arizona Press. 1996. Pp. viii, 202; map; notes; references; index.

¶ *This is a thorough but concise history for control and use of the land of the greater Grand Canyon in regard to the desires of various individuals, groups, and entities. Five different Indian tribes were historically and geographically associated with Grand Canyon. Newcomers wanted land for grazing, timber, mining, and speculation. In the late nineteenth century a movement was begun to protect much of the area as some kind of National Preserve. Early in the twentieth century the idea to harness the energy and power of the Colorado River entered into the contest of use. The struggle continues today. Morehouse explains and discusses all the various ideas and wishes in "A Place Called Grand Canyon."*

163. PEARSON

STILL THE WILD RIVER RUNS. CONGRESS, THE SIERRA CLUB, AND THE FIGHT TO SAVE GRAND CANYON. By Byron E. Pearson. Tucson: The University of Arizona Press. 2002. Pp. xxii, 246; ills.; maps; notes; selected bibliography; index.

¶ *In this well researched history, Pearson presents a full and detailed story of the attempt to build Marble Canyon and Bridge Canyon Dams. If you want to know how – too often at least – government really functions, you should read this book. Conservation organizations, especially the Sierra Club, did all they could to thwart these attempts and were given much credit for halting the projects. Pearson's research indicates they may have been given more credit than they deserved. He shows that, in the end, the politicians and the water politics of the West were the ones that really stopped the dams. It is a fascinating story and much more interesting to read than you might think. In the "Journal of the Southwest," Volume 36 Number 2, Summer 1994, Pearson wrote an article titled "Salvation for Grand Canyon: Congress, the Sierra Club, and the Dam Controversy of 1966-1968." This book is a greatly expanded treatment of that topic.*

164. PETERSON/CRAWFORD

VALUES AND CHOICES IN THE DEVELOPMENT OF THE COLORADO RIVER BASIN. Edited by Dean F. Peterson and A. Berry Crawford. Tucson: The University of Arizona Press. 1978. Pp. xiv, 337; ills.; maps; tables; literature cited; appendixes; index.

¶ *The title says almost all that needs to be said. Twenty-five authors, all very knowledgeable about their topics, have contributed to this volume. Of particular interest are Chapter Five, "Policy Goals and Values in Historical Perspective" by Henry P. Caulfield, Jr.; Chapter Six, "Agriculture and Salinity" by B. Delworth Gardner and Clyde E. Stewart; Chapter Seven, "Energy Resources Development" by Jared Carter; and Chapter Thirteen, "Conservation and Preservation of Aesthetic Values: A Matter of Choice" by Russell Gum.*

165. STEGNER

a

THIS IS DINOSAUR. ECHO PARK COUNTRY AND ITS MAGIC RIVERS. Edited by Wallace Stegner. New York: Alfred A. Knopf. 1955. Pp. 97; ills.; map end papers; the contributors.

b

THIS IS DINOSAUR. ECHO PARK COUNTRY AND ITS MAGIC RIVERS. Edited and with a new foreword by Wallace Stegner. Boulder: Roberts Rinehart, Inc., Publisher. 1985. Pp. x, 93; ills.

¶ *This is a landmark book. It was a "flexing of muscles" by the people opposed to the proposed Echo Park and Split Mountain Dams. Their opposition was based on the fact that these dams would not leave the National Monument "unimpaired" which is the word used in the Act of Congress that outlines the charge to the National Park Service. There are seven chapters; the writers are Wallace Stegner, Eliot Blackwelder, Olaus Murie and John W. Penfold, Robert Lister, Otis "Dock" Marston, David Bradley, and Alfred A. Knopf. They make a very strong case for keeping dams and backwater out of Dinosaur National Monument and all other areas administered by the National Park Service. The text in* b *appears unchanged but* a *has several more photographs, some in color. Some photographs and sketches in* b *are not found in* a. *Frederick R. Rinehart, the publisher, has a note in* b. *Alfred Knopf, the publisher of* a, *donated enough copies to give one to each member of Congress.*

166. TERRELL

WAR FOR THE COLORADO RIVER. By John Upton Terrell with a foreword by Edward Maddin Ainsworth. Glendale: The Arthur H. Clark Company. 1965. 2 volumes; Pp. 325, 323; maps; appendixes; sources and references; indexes.

❡ *Volume One is sub-titled "The California-Arizona Controversy" and Volume Two "Above Lee's Ferry – The Upper Basin." In these two volumes, Terrell tells the full history of the importance of water in the West and the desire of the many individuals and organizations to control that water. He picks up the story with the Colorado River Compact of 1922 and details the fights between special interests in the different states and the fights between the upper and lower basins. The whole story up to the early 1960s is here and much of the text involves the Colorado River Storage Project. These volumes represent a complete piece of research.*

122

167.UNITED STATES DEPARTMENT OF THE INTERIOR

DRAFT GENERAL MANAGEMENT PLAN AND ENVIRONMENTAL IMPACT STATEMENT GRAND CANYON NATIONAL PARK. By the National Park Service. Denver: National Park Service. March, 1995. Pp. ix, 321; maps; ills.; appendixes; index; spiral bound paperback.

¶ *This oversized volume contains a great deal of information about the future of Grand Canyon National Park in relation to the management plan that was to be put in place. Five alternatives were outlined and the steps involved in each were listed and discussed. A lot of thought went into these ideas and the participants justified their recommendations. The quality of the park and therefore quality of the visitor experience in the future is very dependent on the right decisions in regard to the management of the park.*

168. UTAH WILDERNESS COALITION

WILDERNESS AT THE EDGE. A CITIZEN PROPOSAL TO PROTECT UTAH'S CANYONS AND DESERTS. The Utah Wilderness Coalition: Salt Lake City: The Utah Wilderness Coalition with an introduction by Wallace Stegner and a foreword by the Honorable Wayne Owens. 1990. Pp. 400; ills.; maps; appendixes; abbreviations; glossary; acknowledgments; UWC and BLM equivalents; references; index to wilderness areas and units; paperback.

¶ *Many very dedicated people, after spending countless days in the field, contributed to the writing of this large compilation of information. Wilderness areas cannot be saved unless the right people know about them. Remember Glen Canyon! This book records the effort to make more people aware of the areas in Utah that merit protection. The introduction and the foreword put it all in proper perspective.*

169. WEATHERFORD/BROWN

NEW COURSES FOR THE COLORADO RIVER. MAJOR ISSUES FOR THE NEXT CENTURY. Edited by Gary D. Weatherford and F. Lee Brown with a foreword by Bruce Babbitt. Albuquerque: University of New Mexico Press. 1986. Pp. xviii; 253; ills.; maps; notes; appendix; index.

❡ *Many different experts in their fields have contributed to this collection of essays dealing with the history, future use, and management of the Colorado River. The successes since the 1922 Colorado River Compact and the shortcomings for the twenty-first century are outlined. This book, as much or more than any other in the literature, should bring us to the realization that Powell was right about water in the West: "... for there is not sufficient water to supply these lands."*

170. WESTWOOD

ROUGH-WATER MAN. ELWYN BLAKE'S COLORADO RIVER EXPEDITIONS. By Richard E. Westwood with a foreword by Bruce Babbitt. Reno: University of Nevada Press. 1992. Pp. xxi, 259; ills.; maps; epilogue; notes; bibliography; index.

❡ *Not since Stanton has such a detailed mapping of any part of the Colorado River system been attempted. Where Stanton was looking for a place to build a railroad, this group, the United States Geological Survey, was looking for places to build dams. This book tells that story but it is also the biography of Elwyn Blake, a young and, in the beginning, very inexperienced boatman. Working his way down the San Juan on his first trip, he became a much better boatman. As a result of his increased skill in handling a boat under the difficult conditions of surveying, he was the only person to be a member of all three of the mapping expeditions. The rivers included in the mapping surveys were the San Juan in 1921, the upper Green in 1922, and the Colorado through Grand Canyon in 1923. These were not pleasure trips but very difficult and dangerous work.*

PART X

FICTION

171. ABBEY

a

THE MONKEY WRENCH GANG. By Edward Abbey.
Philadelphia and New York: J. B. Lippincott Company.
1975. Pp. 352.

b

THE MONKEY WRENCH GANG. By Edward Abbey with
illustrations by R. Crumb. Salt Lake City: Dream Garden
Press. 1985. Pp. 356; ills.

c

THE MONKEY WRENCH GANG. By Edward Abbey with
illustrations by R. Crumb. Salt Lake City: Dream Garden
Press. 1990. Pp. 356; ills.

❡ *Abbey's "Desert Solitaire" gave the alert for action
needed by environmentalists and the followers of the Earth
First! movement. "The Monkey Wrench Gang" made some
suggestions as to how to go about it. Abbey knew the country
first-hand and used that knowledge to the "gang's" advantage.
The chase will keep you on the edge of your seat, but hold on
tight or you will get thrown off. Abbey didn't always take
himself too seriously and you shouldn't either. In the dark of
night Abbey probably wouldn't push the plunger to detonate
the explosive, but he might hold the flashlight! This is a wild
tale well told.*

The Tenth Anniversary Edition, b, has an added chapter and illustrations. It was issued in a trade edition of 5000 copies in black cloth. Issued simultaneously were fifteen slipcased, lettered copies designated "Publisher's Presentation Copy," and 250 numbered copies, all signed by Abbey with a signed print by Crumb laid in. This is Abbey's only signed limited edition. The 1990 edition, c, is in red cloth and has additional drawings by R. Crumb.

172. HANNON

GLEN CANYON. A NOVEL. By Steve Hannon. Denver: Kokopelli Books. 1997. Pp. ix, 634; ills.; endpaper maps; afterword.

❡ *Hannon has extensive expertise in nuclear weapons, sailing ships, international intrigue, the history of the Ancestral Puebloans, the Colorado River, and more. With a great amount of writing skill, he weaves all these things together to make a very exhilarating story. Edward Abbey would have enjoyed this novel but he would have enjoyed it even more if it had been a documentary! When you have finished the last chapter go back and read chapter one again. It is even more beautiful and moving than when you read it first.*

173. HENRY

BRIGHTY OF THE GRAND CANYON. By Marguerite Henry with illustrations by Wesley Dennis. Chicago: Rand McNally & Company. 1953. Pp. 222; map; ills., acknowledgments.

❡ *A delightful story of a Grand Canyon burro that was befriended by an old prospector. After the prospector passed away Brighty became wild again. He did, however, retain his taste for flapjacks and other camp food and would accept any treats he was offered as long as he wasn't expected to do any work in return. There is much in this book that is fact and Uncle Jimmy Owen and Theodore Roosevelt both play important parts in the story. Most of the feral burros that were in the Canyon have been removed but Brighty can stay forever. This is a story for children of all ages.*

174. VERNON

THE LAST CANYON. By John Vernon. Boston: Houghton Mifflin Company. 2001. Pp. 336; map; epilogue; afterword.

¶ *Vernon weaves two story lines into this work of fiction. One is that of John Wesley Powell and his crew of river runners. The second is of Toab, a Shivwits Paiute, and his little band of relatives. The paths of these two very different groups eventually cross. The story of Powell is familiar and well done but the uniqueness of this book lies in Vernon's ability to craft a very possible and convincing set of circumstances that put the Paiute band north of the lower part of Grand Canyon in the late summer of 1869.*

Part XI

THE ARTISTS

175. BABBITT

COLOR AND LIGHT. THE SOUTHWEST CANVASES OF LOUIS AKIN. By Bruce E. Babbitt, with a foreword by Clay Lockett. Flagstaff: Northland Press. 1973 in an edition of 1750 copies. Pp. xiv, 76; ills.; map; chronology; selected bibliography.

¶ *In 1903 the Santa Fe Railroad offered Louis Akin an opportunity to paint the Hopi Indians as part of an advertising campaign. He took it. When he arrived at the village of Oraibi he rented a room for seventy-five cents per week. He stayed a year, producing at least two dozen paintings. He continued to do work for the Santa Fe and made his first painting of Grand Canyon in 1904, making him one of the early Canyon painters. Probably his most famous work is the 1906 painting of El Tovar on the brink of the South Rim. Of the thirteen paintings reproduced in this publication, five are of the Canyon. Most of the selections are from private collections. Bruce Babbitt has written a good biography of the artist and in doing so identified about 125 paintings done by Akin, more than one-fourth of them of Grand Canyon. It was, without doubt, the artist's favorite subject. In 1988 Northland Press did a soft cover printing on the front cover of which was the Akin painting of Grand Canyon that hangs in the Verkamp Curio Shop at the South Rim. The first edition has Akin's "Navajos Watching Field Sports" on the dust jacket. Internally both printings are the same. There was also a specially bound, signed, limited edition of fifty copies of the first edition.*

176. BELKNAP

GUNNAR WIDFORSS. PAINTER OF THE GRAND CANYON. By Bill Belknap and Frances Spencer Belknap, with a foreword by Don Perceval. Flagstaff: Published for the Museum of Northern Arizona by the Northland Press, 1969. Pp. xx, 86; ills.

❡ *Considered by many to be one of the greatest of all painters of Grand Canyon, Gunnar Widforss was relatively unknown and might have remained so if this book had not been published. Born in Sweden in 1879, he painted scenes in much of Europe before coming to the United States, which he first did in 1905, staying for two years in the eastern states. In 1921, on a trip to the Orient, he came to California and was so impressed with the country that he decided to stay. Due to associations with Ansel Hall, then Park Naturalist in Yosemite, Widforss became acquainted with Stephen Mather, Director of the National Park Service. Mather encouraged him to become the painter of the National Parks. He did, and made many paintings, mostly watercolors, of many scenes in the parks but Grand Canyon was his favorite. The Belknaps have reproduced thirty-one of his paintings, eighteen of which are at the Canyon. Widforss Point on the North Rim is named for "Weedy" and is a fitting tribute to his association and love for Grand Canyon. He is buried in Grand Canyon cemetery. There was a specially bound, signed, limited edition of one hundred copies of this book in slipcase with a sketch of Widforss by George L. Collins laid in.*

177. DAWSON/CRAIGHEAD

THE GRAND CANYON. AN ARTIST'S VIEW. A WALK THROUGH GRAND CANYON NATIONAL PARK. Paintings and drawings by John D. Dawson and story by Charles Craighead. Salt Lake City: Haggis House Publications, Inc. 1996. Pp. 64; ills.; paperback.

These are not just scenes of the Canyon but scenes with the true residents of the Canyon in their native habitat. Mule deer, coyote, cougar, and ringtails are among the many mammals represented. The nuthatch, brown creeper, peregrine falcon, white-throated swift, and the canyon wren are a few of the birds Dawson has included along with many of the native plants. Dawson and Craighead hiked together and the text often gives Craighead's version of what Dawson has drawn or painted. It is a delightful experience to "go with them." You can learn a lot about how to look and how to see the complete scene, not just the Canyon itself. This title would have been right at home in the section on flora and fauna.

178. FERNLUND

WILLIAM HENRY HOLMES AND THE REDISCOVERY OF THE AMERICAN WEST. By Kevin J. Fernlund. Albuquerque: University of New Mexico Press. 2000. Pp. xvii, 300; ills.; map; notes; bibliography; index.

It is sometimes a fine line between illustrator and artist but William Henry Holmes, who started out as an illustrator, certainly became an artist of supreme ability. For the most part his subjects did not move but he often had to wait for long periods of time for clouds or mountain mists to move so he could see his subjects clearly. With a sketch pad and pencil, Holmes created some of the most accurate depictions of the vast geological features of the American West that have ever been done. Just look for a while at his "Grand Cañon at the Foot of the Toroweap, Looking East" or his great "Panorama from Point Sublime," both used in the Dutton "Tertiary History of the Grand Cañon District" [Farquhar #73], and then find a photograph that can match the detail. Fernlund's biography covers all of Holmes's long, varied, and exciting career.

179. FOSTER

Exploring the Grand Canyon. Watercolour Diaries 1988-1989. By Tony Foster with an essay, "Challenge of the Grand Canyon" by James K. Ballinger. Cornwall: Newlyn Orion Galleries, Penzance, in association with Montgomery Gallery, San Francisco. 1989. Pp. 32; 30 plates; biography; exhibition tour 1990; acknowledgments; paperback.

¶ *This is the exhibition tour catalogue for an extraordinarily fine group of watercolor paintings. Tony Foster is an Englishman. His skill as a draftsman will remind you of William Henry Holmes and his skill with watercolors will remind you of Gunnar Widforss and Louis Akin. These are not just exceptional watercolors, but most are from vantage points well below the rim as were some works of Gunnar Widforss before him. All of his paintings were nearly complete when he brought them over the rim and each represents anywhere from a few hours to nineteen days in the Canyon. Dr. Ballinger, Director of the Phoenix Art Museum, has written a five-page essay describing the difficulty of using watercolors in the Canyon and Foster's methods for overcoming those difficulties.*

180. HUSEMAN

Wild River, Timeless Canyon. Balduin Möllhausen's Watercolors of the Colorado. By Ben W. Huseman. Fort Worth: Amon Carter Museum and Tucson: University of Arizona Press. 1995. Pp. viii, 232; ills.; maps; notes; bibliography; catalogue of the watercolors; index.

¶ *The first third of the book contains biographical material about Möllhausen, including his artistic development, along with a very good discussion of the Ives Expedition and his relationship with it. Fifty-one of Möllhausen's watercolors are beautifully reproduced and the catalogue tells something of the location of the scene. Möllhausen's two volume book "Reisen..." [Farquhar #22], is quoted at great length, and until a complete translation of this work is published, this is the nearest most of us will come to reading, in his own words, Möllhausen's experiences with the Ives Expedition.*

181. KINSEY

THE MAJESTY OF THE GRAND CANYON. 150 YEARS IN ART. By Joni L. Kinsey, Arnold Skolnick picture editor and designer, and with a foreword by James E. Babbitt. Cobb: First Glance Books, Inc. 1998. Pp. 160; ills.; epilogue; notes; selected bibliography; index of artists.

¶ *Fifty-nine different artists are represented in this collection, from Egloffstein and Möllhausen to many contemporary artists. There is enough text to give an appreciation of each of the artists and their work. Several quotes are printed in the margins that are appropriate to the painting on the page. These quotes are from the writings of, among others, John Van Dyke, J. B. Priestly, E. L. Kolb, Clarence Dutton, C. A. Higgins, and Joseph Wood Krutch. This is a grand collection of grand art of a Grand Canyon.*

182. WILKINS

a

THOMAS MORAN: ARTIST OF THE MOUNTAINS. By Thurman Wilkins. Norman: University of Oklahoma Press. 1966. Pp. xvi, 315; ills.; preface; appendix; bibliography; index.

b

THOMAS MORAN: ARTIST OF THE MOUNTAINS. By Thurman Wilkins with the help of Caroline Lawson Hinkley, and a foreword by William H. Goetzman. Norman: University of Oklahoma Press. 1998, second edition revised and enlarged. Pp. xxii, 429; ills.; notes; bibliography; index.

¶ *There would probably be some disagreement about who is the second best painter of Grand Canyon, but most would say Thomas Moran was the best. If you don't think so stand about fifteen feet back from his seven-foot by twelve-foot "Chasm of the Colorado" and just look at it for ten or fifteen minutes. You might be convinced. Wilkins has written a complete biography of Moran's career which covered much more than his many trips to the Canyon. Of special interest was Moran's first visit when he accompanied Major Powell to what would become Zion National Park, then to Grand Canyon at Toroweap Point, and finally to the Powell Plateau. The 1998 printing contains new information unavailable for the first edition making it a more complete and accurate biography. There are some changes in the illustrations and in* b *the "Chasm of the Colorado" is printed in color.*

Part XII

FLORA AND FAUNA

183. BROWN/CAROTHERS/JOHNSON

GRAND CANYON BIRDS. HISTORICAL NOTES, NATURAL HISTORY AND ECOLOGY. By Bryan T. Brown, Steven W. Carothers, and R. Roy Johnson. Tucson: The University of Arizona Press. 1987. Pp. xv, 302; ills.; maps; appendix; bibliography; index.

¶ *The first chapter introduces the reader to the early ornithologists that worked at the Canyon. The different life zones, vegetation in them, and species of birds are covered nicely along with seasonal changes and how those affect the species populations. The last half of the book lists the species by family and where they might be found. There are no photographs of the birds in this book so you might want to have something like a Peterson or an Audubon guide for positive identification.*

184. MILLER/YOUNG/GATLIN/RICHARDSON

AMPHIBIANS AND REPTILES OF THE GRAND CANYON NATIONAL PARK. By Donald M. Miller, Robert A. Young, Thomas W. Gatlin, and John A. Richardson. Grand Canyon: Grand Canyon Natural History Association Monograph Number 4. 1982. Pp. vii, 143; ills.; maps; bibliography; index; paperback.

¶ *If you hike or backpack the Canyon trails it would be good to know something about the fellow creatures whose home you are invading – especially this group – because there are a few you might wish to avoid or at least not sleep with. There are many different lizards, all harmless except the Gila Monster, lots of frogs and toads, all harmless; and many different species of snakes, including five different types of rattlesnakes. This book describes them all and provides information on habits and habitats. Read this one before you go, not while you are on the trail looking one of these creatures in the eye, trying to find it in the book.*

185. HOFFMEISTER

a

MAMMALS OF GRAND CANYON. By Donald F. Hoffmeister and illustrated by James Gordon Irving. Urbana: University of Illinois Press. 1971. Pp. 183; ills.; maps; index.

b

MAMMALS OF THE ARIZONA STRIP INCLUDING GRAND CANYON NATIONAL MONUMENT. By Donald F. Hoffmeister and Floyd E. Durham. Flagstaff: Northern Arizona Society of Science and Art, Inc. with contributions from the Museum of Natural History, University of Illinois, Urbana, Illinois. Technical Series No. 11. 1971. Pp. 44; ills.; maps; paperback.

¶ *Man shares the Canyon with many other creatures, or rather we should say the many other creatures share their homes with us. Seventy-four species of mammals that were known to live in or near the Canyon when Hoffmeister completed his research are included in* a. *He gives names, habitat, habits, and the seasons the different mammals might be found in certain parts of the Canyon. The second volume,* b, *does much the same for the Arizona Strip and what was then Grand Canyon National*

Monument (now part of the National Park). When you visit, don't be disappointed if you do not see many animals. Many of them are nocturnal and most of them have learned to avoid us. We have taught them that man is the most dangerous animal of all.

186. MCDOUGALL

GRAND CANYON WILD FLOWERS. By W. B. McDougall, edited by Diony H. Sutherland with drawings by Barton A. Wright. Flagstaff: The Museum of Northern Arizona and the Grand Canyon Natural History Association. 1964. Pp. ix, 259; ills.; index of photographs; glossary; index to family, generic, and common names.

¶ *This volume evolved from the "Checklist of Plants of Grand Canyon National Park," prepared by McDougall and last published in 1947. In this book he describes nearly one-thousand different kinds of plants. It was written not only for the average park visitor, who wishes to learn more about and be able to identify the seed plants they might see, but also for the professional botanist who wants to complete a checklist of the species found in the park. It would seem to achieve both of these purposes quite well.*

187. PHILLIPS/RICHARDSON

GRAND CANYON WILDFLOWERS. By Arthur M. Phillips III, with photography by John A. Richardson. Grand Canyon: Grand Canyon Natural History Association. 1979. Pp. 145; ills.; index; paperback.

¶ *Wildflower blossoms add a special beauty to the Canyon. It may be color, shape, texture, aroma, or the contrast of size with the immensity of the Canyon. Whatever their appeal, this book*

can help you find and identify about one hundred and forty of
the more common species in five different color groups. There
is a 1990 printing in a format the right size and shape to carry
in the field.

188. ROSENBERG/OHMART/HUNTER/ANDERSON

BIRDS OF THE LOWER COLORADO RIVER VALLEY. By
Kenneth V. Rosenberg, Robert D. Ohmart, William C.
Hunter, and Bertin W. Anderson. Tucson: University of
Arizona Press. 1991. Pp. xv, 416; ills.; maps; appendixes;
bibliography; name index; subject index; about the
authors.

¶ There have been significant changes in the habitat of the
lower Colorado due to dams and reservoirs. This has had
an effect on both the numbers and the species of birds. This
comprehensive study covers most everything the professional
ornithologist or the casual bird watcher would want to know.
Species, the location, and the season they might be found are
included.

189. WHITNEY

a

A FIELD GUIDE TO THE GRAND CANYON. By Stephen
Whitney. New York: William Morrow and Company, Inc.
1982. Pp. 320; ills.; maps; selected references with chapters;
plate index; general index.

b

A FIELD GUIDE TO THE GRAND CANYON. By Stephen
Whitney. Seattle: The Mountaineers. 1996. Second edition.
Pp. 269; ills.; maps; suggested references; acknowledgments;
index; paperback.

138

¶ *Whitney has given us a complete field guide including geology, climate, plants and animals, and their distribution. Birds, reptiles, amphibians, fish, butterflies, and more are all in an easy to use format. The first edition has some information on history and trails not included in* b. *There are some minor differences in the maps and in some of the illustrations. Either printing is very useful.*

PART XIII

MISCELLANEOUS

190. ABBEY

a

DESERT SOLITAIRE; A SEASON IN THE WILDERNESS. By Edward Abbey. New York: McGraw-Hill Book Company. 1968. Pp. xiv, 269; ills., with drawings by Peter Parnall.

b

DESERT SOLITAIRE; A SEASON IN THE WILDERNESS. By Edward Abbey. Salt Lake City: Peregrine Smith, Inc. 1981. Pp. xvi, 269; ills., with drawings by Edward Abbey and photographs by Gibbs Smith.

c

DESERT SOLITAIRE; A SEASON IN THE WILDERNESS. By Edward Abbey. Tucson: The University of Arizona Press. 1988. Pp. 255. ills., with illustrations by Lawrence Ormsby.

¶ *Abbey spent two consecutive summers in Arches National Monument (now National Park) as a seasonal ranger and came back later for a third summer. He kept journals of his summers in Arches and used those as well as journals and notes from other experiences in the canyon country to write "Desert Solitaire." In these essays he ranges from Arches and Canyonlands to Glen Canyon and Havasu and senses things of which most visitors are unaware. Abbey has written an eloquent testament to the canyon country on its own terms without the benefit of man's "improvements." This is Abbey's first book of non-fiction and remains the most widely read of his several excellent volumes. This is also the book that helped cause the southwest canyon country environmentalists and wilderness activists to become organized. Edward Abbey wanted to be remembered for his fiction. He will be remembered for his essays. The text in a and b is the same but c has over 200 textual changes and a new preface. Abbey considered c to be the final edition and it is. The drawings in b are by Abbey, making this printing unique.*

191. ABBEY/HYDE

SLICKROCK. THE CANYON COUNTRY OF SOUTHEAST UTAH. Words by Edward Abbey, preface, commentary and photographs by Philip Hyde. Introduction by John G. Mitchell. San Francisco and New York: The Sierra Club. 1971. Pp. 143; ills. The Sierra Club Exhibit Format Series.

¶ *For the most part Abbey and Hyde did not travel together but they visited much of the same country. Abbey's experiences took place from 1944 up through and after the flooding of Glen Canyon. Hyde traveled and photographed the slickrock country gathering material for this book over a seven-year period. For starters Abbey describes a crossing of the Colorado on the Chaffin Ferry at Hite, and a trip – including a flash flood – out the old North Wash trail. He continues with visits to many favorite places in his adopted homeland of southeast and south-central Utah. Hyde's commentary includes the Escalante wilderness, the Waterpocket Fold, and the Canyonlands. Much of the area described and photographed was not protected at the time. Canyonlands National Park now includes the Maze section to the west of the Confluence. The Escalante drainage is now part of the Grand Staircase-Escalante National Monument, and the Waterpocket Fold is now part of Capitol Reef National Park. This is vintage Abbey and will remind you of "Desert Solitaire." Hyde's photographs are a perfect supplement to the text. There is little doubt this volume was one of the influencing factors causing those areas to be set aside and protected.*

192. ANNERINO

RUNNING WILD. THROUGH THE GRAND CANYON ON THE ANCIENT PATH. By John Annerino, foreword by Charles Bowden, and photographs by Christine Keith. Tucson: Harbinger House. 1992. Pp. 205; ills; maps; selected bibliography; paperback.

¶ *This book tells about adventures most of us would probably not wish to duplicate. After a bad fall while rock climbing, Annerino fooled his doctors and was able to run and climb again. His first trial climb after the fall was up a new route on Zoroaster Temple in Grand Canyon. It is not uncommon for avid runners to run from rim to rim on the corridor trails, but for a new challenge Annerino decided to run, among other places, most of the length of Grand Canyon, and do it below the rim. In the process he discovered ancient paths, probably trade routes that were used by the Native Americans hundreds of years ago. Annerino forced himself to the extreme limits of physical and mental endurance and reached his goal. It is an exciting story.*

193. ANDERSON

POLISHING THE JEWEL. AN ADMINISTRATIVE HISTORY OF GRAND CANYON NATIONAL PARK. By Michael F. Anderson. Grand Canyon: Grand Canyon Association. 2000. Monograph Number 11. Pp. ix, 116. ills.; maps; appendixes; notes; index; paperback.

¶ *This is a well researched history of the formation and operation of Grand Canyon National Park from before the area was set aside as Grand Canyon Forest Reserve until the Grand Canyon – Parashant National Monument was formed. The stories of the early miners, settlers, visitors, and the parts they played are all here. Anderson discusses the railroad, the concessionaires, road and trail building, and other visitor services, and the politics involved in setting policy, and places them in their proper perspective. He follows the evolution in philosophy of the National Park Service from the early days to the needs of the millions of visitors of the present day. Many historic photographs add interest to the story.*

194. CHESHER

HEART OF THE DESERT WILD. GRAND STAIRCASE – ESCALANTE NATIONAL MONUMENT. By Greer K. Chesher with photographs by Liz Hymans. Bryce Canyon: Bryce Canyon Natural History Association. 2000. Pp. 104. ills.; maps; selected bibliography; index.

¶ *The uniqueness and beauty of the nearly two million acre Grand Staircase – Escalante National Monument can never be put into one book but this one makes a courageous attempt. The main strength of this volume is its emphasis on the entire ecosystem of not just the monument but all the areas around it. It is in the middle of about eleven million acres of national parks, national monuments, wilderness areas, national recreation areas, and national forests, so it is truly the "Heart of the Desert Wild." Chesher sees, understands, and describes with much accuracy and feeling the everyday happenings in this natural world. The outstanding photographs combined with the text make this book a significant contribution to our understanding of why this area deserved to be set aside and protected.*

195. DOUGLAS

WILDERNESS SOJOURN. NOTES IN THE DESERT SILENCE. By David Douglas. San Francisco: Harper & Row, Publishers. 1987. Pp. x, 102. ills.; sources.

¶ *Douglas never tells us where he is hiking but if you have been there you will know. On this seven-day trip he is alone. In the desert silence, he writes of his feelings which might be compared to the early day prophets going into the desert to find inspiration and renewal. His eloquence includes "the geography of faith" where wilderness can, especially if entered alone, become a very religious experience where we realize our dependence on not just ourselves but on a higher power. This is one more good reason to save those special places.*

196. EASTON/BROWN

LORD OF BEASTS. THE SAGA OF BUFFALO JONES. By Robert Easton and Mackenzie Brown with a foreword by Jack Schaefer. Tucson: The University of Arizona Press. 1961. Pp. xiii, 287. ills.; notes; bibliography; index.

¶ *Although only a few chapters relate directly to Grand Canyon, they are of significant importance. Two chapters depict the history of the present-day bison herd found in the south part of House Rock Valley below the Kaibab Plateau. There are also stories of Zane Grey hunting lions on the Kaibab and chasing wild horses with one of his mentors, Buffalo Jones. These are exciting stories and true.*

197. HASSELL

RAINBOW BRIDGE. AN ILLUSTRATED HISTORY. By Hank Hassell with line drawings and maps by R. Sean Evans. Logan: Utah State University Press. 1999. Pp. ix, 173. ills.; maps; notes; bibliography; photograph credits; index.

¶ *This is a complete story of Rainbow Bridge. Hassell includes the geology to tell us why the bridge is there, the meaning and importance of the bridge through the eyes of the Native Americans, the "discovery" in 1909, early-day tourism, and the environmental battles. Chapter One recounts a hike from the old Rainbow Lodge trailhead, down Cliff Canyon, across Redbud Pass, and down Bridge Creek to Rainbow Bridge. Chapter Eight does the same for a hike in from the north side, the trail that comes in through Surprise Valley and follows, for much of its way, the so called "Discovery Trail" of 1909. If you have hiked these trails you will have special enjoyment of going again with Hassell. He describes the experiences exceptionally well.*

198. HEINIGER

GRAND CANYON. Photographs by Ernst A. Heiniger, foreword by Dr. Joseph Wood Krutch, text by Dr. Hans Boesch, Dr. William A. Weber, Dr. Heini Hediger, and Jean Heiniger. Washington and New York: Robert B. Luce Co., Inc. 1975. [Original copyright Berne: Kummerly and Frey, Geographical Publishers, 1971 with text in German.] This edition translated from the German by Ewald Osers. Pp. 157; ills., end paper maps, maps, including one 3-D topographic map [3-D glasses included] of the heart of the Grand Canyon; commentary on pictures.

¶ *This book evolved from the Walt Disney movie "Grand Canyon," which was filmed to be shown with Ferde Grofe's famous music "Grand Canyon Suite." The four text authors have written on: Geography, Geology, and History; Botany; Fauna; and a year at the Grand Canyon. The botany and fauna sections are especially interesting, bringing out some information not usually found in other books on the Canyon. Jean Heiniger, wife of the photographer, has written of the year they spent at the Canyon making the movie and the pictures that eventually went into this book. Neither had seen the Canyon before they arrived. Altogether this is a very interesting story.*

199. HIRST

a

LIFE IN A NARROW PLACE. THE HAVASUPAI OF THE GRAND CANYON. By Stephen Hirst with photographs by Terry and Lyntha Scott Eiler. New York: David McKay Company. 1976. Pp. xi, 302; ills.; maps; preface; author's note; bibliography; notes; index.

b

HAVSUW 'BAAJA: PEOPLE OF THE BLUE GREEN WATER. By Stephen Hirst with photographs by Lois Hirst. Supai: The Havasupai Tribe. 1985. Pp. ix, 259; ills.; maps; foreword; preface; author's note; notes; bibliography; index.

¶ *At one time the Havasupai used much of the South Rim of Grand Canyon and much of the inner canyon as well. In the late 19th century they were still farming at Indian Gardens. They spent summers in the Canyon and winters on top, a pattern of living that had been followed for many hundreds of years. Then, in 1872, the federal government told the Havasupai they had only the small area in the bottom of Havasu Canyon, something less than 600 acres, for their reservation. They tried repeatedly – and for the most part their pleas fell on deaf ears – to regain the lands on the rim. On January 3, 1975 President Gerald Ford signed the bill that restored land to the Havasupai. Both of these books tell the complete story and have nearly identical texts, but in* b *some of the information may be slightly more accurate. Both printings have very interesting, excellent, but different, photographs.*

146

200. HYDE/JETT

NAVAJO WILDLANDS. "… AS LONG AS THE RIVERS SHALL RUN." Photographs by Philip Hyde and text by Stephen Jett, with a foreword by David Brower, edited by Kenneth Brower. San Francisco: Sierra Club 1967. Pp. 160; ills.; attached map inside back cover; epilogue.

¶ *In his brief introduction, Stephen Jett quotes Howard Gorman, long-time Navajo Tribal Councilman: "Crops can be replanted. Stock can reproduce. So can human beings. But the land is not like these. Once it is taken away, it is gone forever." That is the message this volume of the Sierra Club Exhibit Format Series portrays from cover to cover. The Navajo Country is vast and varied and Philip Hyde's chosen images represent it extremely well. It is obvious the text was written by someone quite familiar with, and very respectful of, both the country and the Diné. Excerpts from the writings of Willa Cather, Oliver La Farge, and several other authors are inserted at intervals in the text. However, the most beautiful passages are from the Navajo Creation Myth and Navajo chants as translated by Washington Matthews and Berard Haile. More poetry should carry a message of this significance and be this beautiful.*

201. JONES

RAMBLINGS BY BOAT AND BOOT IN LAKE POWELL COUNTRY. A PACK FULL OF FASCINATING STORIES & PHOTOGRAPHS BY "MISTER LAKE POWELL" AS HE RAMBLED THROUGH THIRTY YEARS OF OUTDOOR ADVENTURES. By Stan Jones. Page: Sun Country Publications. 1998. Pp. 276; ills; photo credits; paperback.

¶ *Jones writes of people and places, legends and lore, history, and hiking. As Powell reservoir was filling he would park his boat or be dropped off, enter, explore, and photograph side canyons that, until the water rose, were inaccessible. There are stories of fishing, of the House Rock Valley bison herd, of petrified forests, of autographs on stone, of Everett Ruess, of Rainbow Bridge, and many more. Stan Jones is a masterful story teller and this is a book you will enjoy for its many suggestions about where to go or as a source of many very enjoyable arm chair excursions.*

202. MAURER

SOLITUDE & SUNSHINE. IMAGES OF A GRAND CANYON CHILDHOOD. By Stephen G. Maurer based on conversations with William G. Bass. Boulder: Pruett Publishing Company. 1983. Pp. 97; ills.; paperback.

¶ *Although this book is written from conversations with William G. Bass, it turned out to be the nearest thing to a biography of his father, William Wallace Bass, that we have in the literature of Grand Canyon. There are fifty period photographs, several of unknown date or photographer. Some were taken by W. W. Bass and many by Frederic H. Maude. All the photographs are in the area of the Canyon near Bass Camp, on the Bass trails, on the Topacoba Trail, and around Havasu Canyon. The comments by William G. Bass are usually related to the photograph on the facing page. Someone should do a full biography of William Wallace Bass. It would be a very interesting story.*

203. MERTZ

a

PALE INK. TWO ANCIENT RECORDS OF CHINESE EXPLORATION IN AMERICA. By Henriette Mertz. Chicago: Ralph Fletcher Seymore, Publisher. 1953. Pp. 158; maps.

b

PALE INK. TWO ANCIENT RECORDS OF CHINESE EXPLORATION
IN AMERICA. By Henriette Mertz. Chicago: Swallow Press
Incorporated. 1972. Second revised edition. Pp. xiv, 175;
ills.; endpaper maps; maps; bibliography; index.

¶ *Mertz uses two ancient Chinese documents, one from nearly
four thousand and the other from about fifteen hundred years
ago, to suggest people from China visited Central and North
America that long ago. From the interpretations that have
been made of the translations of these early documents, there
is enough evidence for a reasonable discussion of the possibility.
One place name was translated to mean the "Great Luminous
Canyon" and another one meant "the place where the sun was
born." This is a story that certainly belongs in the literature
of Grand Canyon. The text is revised in* b *to include some
new material as well as some minor rewriting. The maps are
different but in both they are adequate. There are photographs
in* b *that compare such things as pottery, sculpture, and
architecture from China to those same kinds of items in Central
America. After reading this title you may or may not agree that
the Chinese were here that early, but do you know of another
"Great Luminous Canyon"?*

204. POWELL

SEEING THINGS WHOLE. THE ESSENTIAL JOHN WESLEY
POWELL. By John Wesley Powell, edited by William
deBuys. Washington: Island Press/Shearwater Books. 2001.
Pp. xiii, 388; maps; ills.; works by John Wesley Powell;
secondary works cited; index.

¶ *William deBuys has given us a partial anthology of the
writings of John Wesley Powell, selected as some of the most
significant from the very prolific Powell pen. He has taken
both popular, well circulated writings and some of the more*

149

obscure ones, such as a letter from the "Mouth of Niutah"
[Uinta]. He introduces each of them at length, analyzes them,
and puts them in historical perspective in relation to the needs
of the West of today. For example, many people have heard
of Powell's "Report on the Lands of the Arid Region," though
relatively few have read it. Even fewer know that it was not
the last word from Powell on this topic. In 1890 "Century
Magazine" published three articles which were Powell's "final
edition" on the arid region His thinking had changed some
since 1878. From deBuys's selections and commentary it
becomes very obvious that, even though Powell's writings were
done over one hundred years ago, he is still a wise counselor
for understanding our western lands. This book lets us know
Powell much better and we can better realize what a visionary
thinker he was.

205. PURVIS

THE ACE IN THE HOLE. A BRIEF HISTORY OF COMPANY 818 OF
THE CIVILIAN CONSERVATION CORPS. By Louis Lester Purvis.
Columbus: Brentwood Christian Press. 1989. Pp. 143; ills.;
bibliography; roster of personnel; paperback.

¶ *Soon after the stock market crash of 1929, President*
Franklin D. Roosevelt authorized the Civilian Conservation
Corps, or ccc, as an important part of the national recovery
program. Purvis deals primarily with the enrollees of Company
818 that were assigned to work at Grand Canyon. Among
other things, this group built the River Trail, the Clear Creek
Trail, and the trans-canyon telephone line. Purvis was one of
the members of Company 818 and by using records on file at
Grand Canyon, interviews with other members, and personal
experience, he gives us a record of a significant but relatively
unknown chapter in the Canyon's history.

206. PYNE

FIRE ON THE RIM. A FIREFIGHTER'S SEASON AT THE GRAND CANYON. By Stephen J. Pyne. New York: Weidenfeld & Nicolson. 1989. Pp. ix, 323; endpaper maps; afterword; glossary.

¶ *A well told story of the activities of fire fighters in the western forests. Pyne spent fifteen seasons on the North Rim and in his fourth year became foreman of a fire crew. He describes with first-hand narrative the boredom, the fear, the excitement, the friendships, and the beauty that he experienced. Some of the discussion involves the park bureaucracy, which at times runs at cross purposes to the needs of the environment and the stated goals of the National Park Service.*

207. REA

CANYON INTERLUDES. BETWEEN WHITE WATER & RED ROCK. By Paul W. Rea. Salt Lake City: Signature Books. 1996. Pp. xiv, 280; paperback.

¶ *This is a group of essays by a writer who is very sensitive to the land. For twenty years Rea has hiked the canyons and rafted the rivers of the Colorado Plateau. He takes you to many special places and relates the meaning, for him, of those places. If you have already been to some of these sites you will relive your experience from your own special memories. If you haven't been, take this book with you and read it while you are there.*

208. RINGHOLZ

a

URANIUM FRENZY. BOOM AND BUST ON THE COLORADO PLATEAU. By Raye C. Ringholz. New York: W. W. Norton & Company. 1989. Pp. 310; ills., maps; epilogue; notes; index.

b

URANIUM FRENZY. SAGA OF THE NUCLEAR WEST. By Raye
C. Ringholz. Logan: Utah State University Press. 2002,
revised and expanded. Pp. 344; ills.; maps; epilogue; notes;
index; paperback.

❡ *Using Charlie Steen, who found the Mi Vida mine, as a*
main character in the story of prospecting for uranium on the
Colorado Plateau, Ringholz writes an exciting account of this
period in the Four Corners country. Like the gold and silver
seekers of an earlier time, the uranium prospectors came in by
the thousands; and like the hunters of gold and silver, most
went home poorer than when they came. The prospectors
were, for the most part, the builders of the present jeep trails,
especially in southeastern Utah, where the sleepy little town
of Moab became the "Uranium Capital of the World." In the
1950s the miners did not understand the dangers of radiation
associated with uranium mining and Ringholz describes
several case histories. This is a well documented story and an
important part of the history of the Colorado Plateau. The
2002 printing has five more chapters than the first edition and
brings several things up to date.

209. ROBERTS

IN SEARCH OF THE OLD ONES. EXPLORING THE ANASAZI
WORLD OF THE SOUTHWEST. By David Roberts. New
York: Simon & Schuster. 1996. Pp. 271; ills.; map;
acknowledgments;   appendix;   glossary;   selected
bibliography; index.

❡ *The geography is quite different in this book than in the book*
David Roberts co-authored with Bradford Washington about
Mt. McKinley – but it is equally well written. Roberts was part
of the first group to climb the Wickersham Wall. In this book
he visits Mesa Verde, Grand Gulch, Moqui Canyon, Cedar

Mesa, and some other places that are hard to find on a map
if you can find them at all. Most of the places he goes in this
book have prehistoric ruins where either the Anasazi or the
Fremont cultures lived. He was guided by Fred Blackburn on
several of his trips, and from Blackburn he learned about the
Outdoor Museum. Everyone who hikes and backpacks the
plateau country needs to know the philosophy of the Outdoor
Museum. It is an idea whose time has come.

210. RUSHO

a

EVERETT RUESS. A VAGABOND FOR BEAUTY. By W. L. Rusho
with an introduction by John Nichols and an afterword
by Edward Abbey. Salt Lake City: Gibbs M. Smith, Inc.,
Peregrine Smith Books. 1983. Pp. xi, 228; ills.; maps; notes;
Everett Ruess – a chronology; index.

b

WILDERNESS JOURNALS OF EVERETT RUESS. Edited and with
an introduction by W. L. Rusho and a foreword by Gibbs M.
Smith. Salt Lake City: Gibbs Smith, Publisher. 1998. Pp.
216; ills; index; paperback.

c

EVERETT RUESS. A VAGABOND FOR BEAUTY & WILDERNESS
JOURNALS. The first title was edited by W. L. Rusho with an
introduction by John Nichols and an afterword by Edward
Abbey. The *Wilderness Journals,* by Everett Ruess were
edited by W. L. Rusho with a foreword by Gibbs M. Smith.
Salt Lake City: Gibbs Smith, Publisher. 2002. Pp. 441; ills.;
map; notes; a chronology.

¶ *In November 1934 Everett Ruess camped in Davis Gulch,*
an Escalante River tributary not far above the Colorado River
in Glen Canyon. He was not quite twenty-one years of age
and has not been seen since. For a *Rusho has collected all the*

available letters to parents and friends that Ruess sent, many of which were printed in Farquhar #119. Rusho also gathered newspaper items about Ruess's disappearance and the searches that were made. Rusho interviewed everyone he could find that had known Ruess, and talked at length with people familiar with the country. He also has spent considerable time in the country that Ruess loved. Rusho has put together a complete look at this very unusual and talented desert wanderer. It is a compelling story about a young man who might have agreed with Henry David Thoreau who wrote "Not till we are lost, in other words not till we have lost the world, do we begin to find ourselves, and realize where we are and the infinite extent of our relations."

The "Wilderness Journals," b, contain Everett's journal for 1932 in southern and northeastern Arizona, New Mexico, and Colorado, and his California journal for 1933. The texts in c are the same as they were in both the previously published volumes. The photographs are basically the same but most are enlarged.

211: SCHWARTZ

ON THE EDGE OF SPLENDOR. EXPLORING GRAND CANYON'S HUMAN PAST. By Douglas W. Schwartz. Santa Fe: The School of American Research. nd, circa 1989. Pp. 80; ills.; maps; references; paperback.

¶ *This book is a good introduction and overview to the study of Grand Canyon archaeology, including brief discussions of the prehistoric and historic cultures and their association with the Grand Canyon area. Schwartz goes into enough depth to stimulate your interest and the references at the back are a place to begin study further. There are literally hundreds of books on the archaeology of the Southwest. One has to wonder if the early residents realized and appreciated the beauty that surrounded them "on the edge of splendor."*

154

212. SHORT

RAGING RIVER – LONELY TRAIL. TALES TOLD BY THE CAMPFIRE'S GLOW. By Vaughn Short and illustrated by Joanna Coleman. Tuscon: Two Horses Press. 1978. Pp. 156; ills.; map.

¶ *As the subtitle indicates, these are campfire tales but they are told in verse. A history of relevant events introduces each of the more than fifty poems. Quoting from the dust jacket, "The smoke of many campfires mingle to make this book. Campfires built in the deep, red canyons of Utah, by the big, roaring rapids of the mighty Colorado River and high on the rising, blue mountains of Southern Arizona." One reviewer called Vaughn Short "… the Robert Service of the Canyon Country." That puts Short in a rather elite class.*

213. SKINNER

ONLY THE RIVER RUNS EASY. A HISTORICAL PORTRAIT OF THE UPPER GREEN RIVER VALLEY. By H. L. Skinner. Boulder: Pruett Publishing Company. 1985. Pp. 131; ills.; map; notes on sources; paperback.

¶ *The upper end of the Green River – the longest branch of the Colorado River system – is the theme of this book. About the only things this valley has in common with the upper valley of the shorter branch – the upper Colorado – is that a river runs through it. The upper Green runs through natural lakes, and except for a small National Forest Service campground, has no tourist facilities for many miles. A few ranch buildings and corrals and a dearth of roads allow the area to look much like it did when the mountain men first trapped beaver on the river they called the "Shetskadee." This book is a good overall geological and human history of the upper Green River.*

214. STEGNER

The Sound of Mountain Water. The Changing American West. By Wallace Stegner. Garden City: Doubleday & Company, Inc. 1969. Pp. 286.

❡ *Most if not all these sixteen essays were published previously but it is good to have them all together, as Frank Dobie would say, "… between hard covers." Not all of them pertain to the Colorado River or Plateau, but all are about the West. There is a trip down the San Juan and Glen Canyon and a chapter on the flooding of Glen Canyon with back-water from the dam. There is a ride down the Topacoba Trail to Supai that you won't soon forget, nor likely wish to duplicate. Stegner's often quoted "Coda: Wilderness Letter," written in 1960, is here. Reading it today we can appreciate it even more than when it was fresh more than forty years ago. If you, too, like the sound of mountain water, you will want to read the "Overture" more than once.*

215. SURAN

The Kolb Brothers of Grand Canyon. Being a Collection of Tales of High Adventure, Memorable Incidents, & Humorous Anecdotes. By William C. Suran. Grand Canyon: Grand Canyon Natural History Association. 1991. Pp. 60; ills.; further reading; paperback.

❡ *Ellsworth Kolb arrived at Grand Canyon in the fall of 1901 and there was a Kolb presence on the South Rim until Emery's death in 1976. Quoting from Suran's preface, "I have included here only a few of the most fascinating tales." Many Kolb photographs are here including some from the 1911-12 river trip as well as from other explorations in the Canyon.*

Suran has written a full length manuscript on the life of the Kolb Brothers titled "With the Wings of an Angel: A Biography of Ellsworth and Emery Kolb, Photographers of Grand Canyon" but has not yet found a publisher. It is available on the website of the Grand Canyon Historical Society.

216. TAYLOR

GRAND CANYON'S LONG EARED TAXI. By Karen L. Taylor. Grand Canyon: Grand Canyon Natural History Association. 1992. Pp. vii, 52; ills.; map; credits; paperback.

❡ *How many people have ridden how many mules how many miles in Grand Canyon? We might as well ask how many stars are in the night sky over Grand Canyon. This little book covers the history of the early trails and the use of mules on them by the likes of John Hance, Ralph Cameron, Bill Bass, and Louis Boucher. Taylor discusses choosing the right type of mule for tourists to ride, and the training and care of that mule. There is a lot of history here: of mules, of trails, and of people at Grand Canyon.*

217. UTAH HISTORICAL SOCIETY

a

THE COLORADO... RIVER OF THE WEST.

"The Colorado River – The Physical and Biological Setting" by Angus M. Woodbury. "The Aboriginal Peoples" by Jesse D. Jennings. "Powell of the Colorado" by William Culp Darrah. "Reclamation and the Colorado" by Jay R. Bingham. "Green River: Main Stem of the Colorado" by William M. Purdy. "The Engineer and the Canyon" by Dwight L. Smith. "Historic Glen Canyon" by C. Gregory Crampton. "River Runners: Fast Water Navigation" by Otis Marston. "Land of Space Enough" by Jack Goodman. Introduction by George D. Clyde in Utah Historical Quarterly, Vol. 28, No. 3., Salt Lake City: Utah Historical Society. July, 1960. Pp. 195-324. ills.; maps; paperback.

b

JOHN WESLEY POWELL AND THE COLORADO RIVER CENTENNIAL EDITION.

"John Wesley Powell and an Understanding of the West" by William C. Darrah. "John Wesley Powell, Anthropologist" by Don D. Fowler and Catherine S. Fowler. "The Lost Journal of John Colton Sumner" by O. Dock Marston. "John Wesley Powell, The Irrigation Survey, and the Inauguration of the Second Phase of Irrigation Development in Utah" by Thomas G. Alexander. "Francis Bishop's 1871 River maps" edited by W. L. Rusho. "F. S. Dellenbaugh of the Colorado: Some Letters Pertaining To The Powell Voyages and the History of the Colorado River" edited by C. Gregory Crampton. "How Deadly Is Big Red?" by P. T. Reilly. "The Powell Survey Kanab Base Line" by Robert W. Olsen, Jr. "River Running 1921: The Diary of E. L. Kolb" edited by W. L. Rusho. In Utah Historical Quarterly, Vol. 37, No. 2. Salt Lake City: Utah Historical Society. Spring, 1969. Pp. 146-283. ills.; maps; paperback.

c

COLORADO RIVER COUNTRY.

"Before Powell: Exploration of the Colorado River" by Melvin T. Smith. "Charles Kelly's Glen Canyon Ventures and Adventures" by Gary Topping. "The Bernheimer Explorations in Forbidding Canyon" by Harvey Leake and Gary Topping. "Les Voyageurs Sans Trace: The Decolmont–Deseyne Kayak Party of 1938" by Roy Webb. "Norman Nevills: Whitewater Man of the West" by P. T. Reilly. In Utah Historical Quarterly, Vol. 55, No. 2. Salt Lake City: Utah Historical Society. Spring, 1987. Pp. 103-200; ills.; maps; paperback.

¶ *These three issues deal exclusively with articles about the Colorado River and the Colorado Plateau. Each issue contains much information not found in other sources.*

218. WILKINSON

Fire on the Plateau. Conflict and Endurance in the American Southwest. By Charles Wilkinson with illustrations by Diane Sylvain. Washington: Island Press/ Shearwater Books. 1999. Pp. xiv, 402; ills.; end paper maps; maps; notes; acknowledgments; index.

¶ *Here is a complex story which investigates the history of the Colorado Plateau from exploration and exploitation to environmental awareness and concern. The conflicts between the Native Americans and the successive waves of trappers, miners, explorers, and settlers are among the main themes of the book. Wilkinson, employed early in his law career with the Native American Rights Fund, worked with several different groups of Native Americans, but especially with the Hopi and Navajo. These experiences influenced his philosophy and his future. If you have a problem with your blood pressure you may want to avoid the chapters that deal with Peabody Coal Company, a Salt Lake City lawyer, and the Hopi – if you believe in honesty, ethics, and fair play in business, the well documented information about this trio will surely raise it.*

219. WILLIAMS

Pieces of White Shell. A Journey to Navajoland. By Terry Tempest Williams with illustrations by Clifford Brycelea. New York: Charles Scribner's Sons. 1983, 1984. Pp. 162; map; ills.; acknowledgments; notes; bibliography.

¶ *On a leaf near the front of this book is a quote from Albert Einstein: "The most beautiful and most profound emotion one can experience is the sensation of the mystical ... It is the source of all true science." This thought is the essence of these perceptive and sensitive stories. Williams, well traveled in Navajoland and well acquainted with the Navajo People, is a skilled storyteller of her own Mormon family history as well as of the stories from the Navajo culture. She finds several parallels and relates them in this intuitive look into another culture – a culture whose way of life is difficult for people of European origin to understand. It is a beautifully conceived and beautifully written collection of glimpses into another world of thought.*

P_{ART} XIV

PERIODICALS

Over the years hundreds of different periodicals have carried articles about Grand Canyon, the Colorado River, and the Colorado Plateau. Many of them are well written and of great interest and some of the information is not found in any other printed source. Earle Spamer has listed all of those he could find in his "Bibliography of the Grand Canyon." There are a few periodicals being published today that deal specifically with the Canyon, the Rivers, and the Plateau and because of that they merit being pointed out. These also contain much information the reader will not find in any other printed source. Listed below are the publisher and the titles of the periodicals and something about their history and content.

220. COLORADO PLATEAU RIVER GUIDES

The Confluence

Published by the Colorado Plateau River Guides, Moab, Utah. Volume 1, Number 1, Winter, 1994. Currently from two to four issue numbers are being published each year.

¶ *This periodical is especially produced for the guides that work the Yampa, San Juan, Green, and Colorado Rivers above the Confluence, and on down through Cataract Canyon to Powell reservoir. Along with river-related information, each volume contains several articles that deal with natural and human history, from the early explorations on the rivers and the plateau, up to events still happening today.*

221. DAN O'LAURIE CANYON COUNTRY MUSEUM

CANYON LEGACY

Published by the Dan O'Laurie Canyon Country Museum, Moab, Utah. Volume 1, Number 1, Spring, 1989. Currently four issue numbers are being published each year.

¶ *Every imaginable topic relating to the Colorado Plateau and the Canyon Country may be included in this periodical. Many articles are about early history in the Canyon Country while some are about current topics. Photographs are usually included, many of them historic. All authors are very knowledgeable in the area about which they write.*

222. GRAND CANYON HISTORICAL SOCIETY

a

G. C. P. S. NEWS LETTER

Published by the Grand Canyon Pioneers Society, Grand Canyon, Arizona. Volume 1, Number 1. May, 1990. Three issue numbers were published under this title.

b

GRAND CANYON PIONEERS SOCIETY NEWSLETTER

Published by the Grand Canyon Pioneers Society, Grand Canyon, Arizona from Volume 2, Number 1 through Volume 4, Number 6. Thereafter the address is Flagstaff, Arizona. Twenty-five issue numbers were published under this title.

c

O'PIONEERS

Published by the Grand Canyon Pioneer Society, Flagstaff, Arizona. Volume 5, Number 7, July 1994. Twenty-one issue numbers were published under this title.

d

THE OL' PIONEER

Published by the Grand Canyon Pioneer Society, Flagstaff, Arizona. Volume 7, Number 4, April, 1996. Currently four issue numbers are being published each year.

¶ *As the various titles for this newsletter (now an enjoyable and informative periodical) indicate, it contains information about people and events from an earlier time. At the time this organization was formed, many of the members had lived and worked at the Canyon for many years. Shared stories of their experiences at the Canyon were printed in their newsletter along with coming events. As interest spread to "outsiders" many wanted to join and were welcomed by "the pioneers," and now there are many members who have never lived or worked at the Canyon, but share the same interest and feeling for it. This is one of the reasons that the organization has recently changed its name to the Grand Canyon Historical Society.*

223. GRAND CANYON RIVER GUIDES

a

GRAND CANYON RIVER GUIDES

Published by the Grand Canyon River Guides, Flagstaff, Arizona. Volume 1, Number 1, April, 1988. Nine issue numbers were published under this title.

b

The News

Published by the Grand Canyon River Guides, Flagstaff, Arizona. Volume 4, Number 2, Summer, 1991. Twelve issue numbers were published under this title.

c

Boatman's Quarterly Review

Published by the Grand Canyon River Guides, Flagstaff, Arizona. Volume 7, Number 2, Spring 1994. This title is still being used and four issue numbers are being published each year.

¶ *Several of the articles in each issue contain information of interest or information needed by many of the people who work on the River. Many issues have information about the geology of the Canyon and about the changing of the rapids in the river. Each issue has articles for anyone interested in the history of river running. Quite often there is an interview with a boatman, often times retired, relating their personal river experiences.*

224. GRAND CANYON TRUST

Colorado Plateau Advocate

Published by The Grand Canyon Trust, Flagstaff, Arizona. Volume 1, Number 1, August, 1989. This title is still being used and three to four issues are being published each year but with a month or season designation rather than a volume and number designation.

¶ *With this publication, the Grand Canyon Trust informs members and other readers about issues and solutions that affect any part of the Colorado Plateau, especially Grand Canyon. Included are issues such as returning the natural quiet to Grand Canyon, restoring the Colorado River to something close to a natural condition, advocating for clean air across the*

*Colorado Plateau, and finding the best ways to work with all
agencies that manage areas on the Colorado Plateau.*

225. MUSEUM OF NORTHERN ARIZONA

a

PLATEAU

Published by the Museum of Northern Arizona

Volume 12, Number 1, July, 1939 through Volume 65, Number 4, 1994. In the mid-1970s the format was enlarged, a slick color pictorial cover was added, and many color photographs were used with the articles. During the years the title *Plateau* was in use, four issue numbers were printed each year.

b

CAÑON JOURNAL

Published by the Museum of Northern Arizona, Flagstaff, Arizona and The Grand Canyon Association, Grand Canyon, Arizona. Volume 1, Number 1, Spring, 1995 through Volume 2, Number 2, Fall/Winter, 1996. During the years the title *Cañon Journal* was in use two issue numbers were printed each year.

c

PLATEAU JOURNAL

Published by the Museum of Northern Arizona, Flagstaff, Arizona. What would be Volume 1, Number 1 is designated Summer, 1997. Two issue numbers are being printed each year.

¶ *The present periodical being published by the Museum of Northern Arizona has evolved from very distinguished ancestors. No matter what title it carried this has always been a very scholarly periodical. It deals, as one would expect from the different titles, only with topics pertaining to the Colorado Plateau – but those can be anywhere between anthropology and zoology. The list of authors over the years is very impressive. The names run from Abbey, Bartlett and Colton to Wetherill, Wyman and Zwinger.*

Beginning in July of 1928, Volume 1, Number 1 was published with the title "Notes from the Museum of Northern Arizona." In September 1929, Volume 2, Number 3 was issued with the title "Museum Notes of the Museum of Northern Arizona." In July 1939, Volume 12, Number 1 was issued with the title changed to "Plateau." This title was used until 1995 when a joint venture between the Museum of Northern Arizona and the Grand Canyon Association published Volume 1, Number 1 of "Canyon Journal." The title was changed to "Plateau Journal" for the Summer 1997 issue and it again began with Volume 1, Number 1 although it was just called Summer 1997. The Winter 1997/1998 issue was marked Volume 1, Number 2. Beginning with Volume 5, No. 1, Spring/Summer 2001, the Museum of Northern Arizona was the sole publisher and the title was still "Plateau Journal." The Grand Canyon Association and many other groups and associations were listed as Plateau Partners.

INDEX OF TITLES, AUTHORS, AUTHORS OF INTRODUCTIONS & FOREWORDS, EDITORS, PHOTOGRAPHERS, AND ILLUSTRATORS

(The index refers to the item number.)

176

Based on a design by
WARD RITCHIE
1953

Printing limited to
150* clothbound
400 trade paper

*50 signed and numbered
in slipcase with
the 2003 reprint of
*The Books of the Colorado
River & the Grand Canyon*
by Francis P. Farquhar